FRUITS OF CARE

A User's Guide to Family Caregiving

GAEL CHIARELLA ALBA
SHARON ZINT MARTS

ISBN: 978-1-6847-0675-4 (sc)
ISBN: 978-1-6847-0676-1 (e)

Library of Congress Control Number: 2019908533

Front Cover Art Concept by Rachel K. Hiles

Back Cover Art by William L. Zint, Jr.

Illustrated photos by Gael Chiarella Alba

Lulu Publishing Services rev. date: 08/12/2019

CONTENTS

PREFACE

We were a natural fit for becoming coauthors of this book! Fate brought us together, yet distance made writing this book a challenge. When we met, we both had come to a place where our personal lives as caregivers and our career backgrounds allowed us to collaborate on something we knew would benefit others. We knew we wanted to share our stories and help others find sense and meaning for their own paths. It was a way to sow fruits from the seeds that took root and grew in our lives.

When we shared what we wanted to write, people said, "How can I learn more?" In our book, we wanted to explain the caregiver experience—help people to see they are not alone—and give advice, promote resilience, and allow for places to see joy as caregivers. We know how compassion and self-awareness are an essential component of the caregiver journey, and we felt guiding questions at the end of each chapter in the book could engender the reader's own stories to be told on paper. It is precisely through reflection—looking at the past, present, and future—that there can be detachment from the immediate, and perhaps a pause will allow you, the reader, to feel more in control and present for the hard realities *and* the moments of wonder and gratitude during this experience of family care.

The reflections at the end of each chapter fall into four categories: clarity, action, growth, and introspection. By taking the time to read and respond to the writing prompts in each chapter, you can create time for introspection, gain clarifying insights, and enjoy personal growth that will inform your actions in various situations.

We hope the term *caregiver* extends its meaning for you in new ways that help you see how the fruits of your labor are worth your efforts in more ways than you can see.

INTRODUCTION

There are only four kinds of people in the world—
Those who have been caregivers,
those who are currently caregivers,
those who will be caregivers,
those who need caregivers.
—Rosalynn Carter

When life revolves around upholding another's health and well-being, you may find yourself as a caregiver whether you call yourself one or not. We use the term "family caregiver" to signify any caregiver, carer or care partner who is an unpaid or paid relative or friend who helps an individual with his or her activities of daily living. The caregiver may be prefixed with "spousal", "child", "parent", "young" or "adult" to signify the type of relationship, but we use the term family caregiver to include them all. This is distinguished as separate from the paid version of a caregiver, or a hired Personal Care Assistant. This is often a difficult path that diverges from your peers. We, the authors, were caregivers in different ways and at different times of our lives—Sharon with a son who had a developmental delay, Gael with care of a family member who had a traumatic injury, both of us with eldercare in various forms.

Since both of us shared this experience of caregiving, we came together to try to figure out why caregiving presented such a unique set of hurdles. Beyond the frustrations of medical challenges, and impatience from those we were caring for, we were also struck by the isolation we ourselves felt during the experience. When Sharon visited support groups as a speaker, she often heard the caregivers use the phrase "we feel invisible." This typified that same sense of isolation and loneliness that affects people who are never alone and are just going through an experience without others who truly understand and see them.

Before we were coauthors together, we sought out like-minded people who were also caregivers to find solidarity, and we also found ways to successfully cope. Sharon studied at graduate school with a focus upon caregiving, and Gael was driven to graduate studies in leadership and organizational development while getting certified in caregiving consulting in order to make sense of the subject. In our own worlds—without knowing one another—we tracked what worked for us.

When we met and found our similar passion for the unique challenges of the family caregiver, we decided to write a book to share our findings with readers and personal stories that epitomize care in families and the unforeseen ups and downs. Although there are many hard aspects of care, for this book, we have a higher goal: we want to show caregivers that it is possible to incorporate joy into life as a caregiver. It does involve a shift in orientation but it is possible. Cultivating this sense of wonder and openness that embody joy is a game changer for caregivers and something we hope our readers will strive to create in their lives.

The Framework of This Book

Although there is never a perfect time to be a caregiver, you may find solace in knowing that you are not alone. Those caring for family members are able to stand up and be counted in ways that were not seen in the past. Census data tells us you are nowhere near alone because people are living longer and need care from a spouse, and there are many adult children who care for elders. Family care can also mean the care of those younger than us like children, teens, or adults who are dependent for a variety of reasons.

Whether care is long term or for a duration that simply seems long, it can be trying. Actions of caregivers might even seem unnoticed, unthanked, and taken for granted. Whether you bought this book because you felt this way or because you are overwhelmed and need a boost, or perhaps just want to be a better family caregiver, you will find it does indeed contain several fruits that you can see again in your life, or uncover anew amidst your actions of caring for others in your family.

Patterns of Care and the Shifts Time Creates as Older Generations Age

We expect care in life—we anxiously await a birth at the end of a pregnancy—whether it be for a child, a grandchild, or another relative. Care in this case is a predetermined fact. Yet many see care in the reverse direction—the care of elders—as a burden, as though it was a surprise that one day with all of America's medical advances, people could still age and need to be cared for by others in their families. This is expected, and it should be embraced.

Think back to the grandparents you remember. As a youth, you probably saw them welcoming your presence, watching your milestones alongside your parents. Then, one day, you noticed a shift in the currents of care. Those who initiated, looked after, and were relied upon in a time of need were not as fast to respond or were wary of getting too involved in the care of another when looking after themselves is harder to do. Perhaps a parent or grandparent suffered a fall or a diagnosis that brought to light their mortality and the shift took place when those who cared are now cared for.

In the generations living before the twentieth century's senior living communities, care of elders was expected to be in the adult child's home. Some cultures even designate which family members are to be the primary caregiver, and others do not. Yet the inevitability of care for others related to us is always there as an underlying way, and how we choose to see and embrace that role is why this book has been written.

Unexpected Care

A child born with a developmental delay, a disability, or a diagnosis found later in youth creates a dimension of unexpected care that one may not have foreseen. Becoming a parent when your child has a special needs diagnosis may require asking questions and taking actions that may feel uncomfortable at first. Pediatricians, specialists, and people within your circles of life may even offer conflicting advice which can be daunting at a time when learning to become an advocate is new to you.

There are also those who are not new parents but rather become caregivers of their own family for such circumstances when a person in the family suffers from a life-changing event that shifts the course ahead— such as a traumatic brain injury during a military deployment, a car accident, a stroke, or another life-altering change. Regardless of how a person becomes a family caregiver, a new normal will find its way, but the path getting there may be fraught with challenges as old and new expectations compete, and a *different* capability will emerge.

This Book's Goals

Gael Chiarella Alba and Sharon Zint Marts wrote this book with all family caregivers in mind, and rather than focusing upon one type of caregiver and the specific situation influencing their care, we hope to find commonality in the experience of finding meaning in care and places where grace and solace exist. Through reflective questions at the end of each chapter, we ask our readers to look inwardly to your life with probing questions that move beyond the daily actions and hint at the hopes and expectations you bring to your care. We will also further your thinking by asking about your understanding of your own capabilities as a caregiver and those you care for. You'll learn advocacy and boundaries and see your life through a different lens that will impart a renewed resilience and strength you can take back into your caregiving role. We will ask that you shift ideas about spearheading care in an autonomous way and see how spreading out care across the system of other family, health care professionals, and even paid staff if needed, can provide rest and necessary respite and isn't anything more than a lifeline you deserve when warranted.

There are books and films in popular culture in which a person looks at their life through a different lens. *A Christmas Carol* (1843) by Charles Dickens shows us past, present, and possible futures that could come to pass with different outcomes. Frank Capra's classic film *It's a Wonderful Life* (1946) is a celebrated Christmas film that depicts how good deeds in one's life only come to be valued when a man is about to take his own life—only to have the story take a turn with an angel who intervenes to change the future. For a humorous take on finding meaning in life, think of Bill Murray's role in *Groundhog Day* (1993) in which his character is "stuck" in the same motions until he learns a better way to overcome his present circumstances with grace. The knowledge each story's protagonists gain as they see their lives differently helps them take actions they might not have foreseen initially. In each case, the characters take heed from events and begin to learn from challenges to empower their own lives in significant ways. Our book is meant to be a way to look at your life from a past, present, and hopeful future. You will be given reflective exercises to complete so that you can learn from vulnerability along with your strengths. By letting down your defenses and then regaining them, a renewed spirit will come with it.

Care will shape our American future—whether we like it or not—as our elders will be living longer life spans, with three and four generations of family enmeshed in care of one another. Care is not always one way with the independent caring for the dependent with words and actions, which may be seen by some as uncomfortable or hard. Perhaps in our new understanding of care in the millennium, we will need to redefine our terms because it is possible that emotional, spiritual, physical, and financial care could come from any generation—upward (as in an adult child who might be taking physical care of elders) or downward (the care by parents or grandparents of children / dependent adult children who may be infants, adolescents, or dependent young adults), or even be a concurrent type of care with one elder taking care of another of a similar adult age.

It is our hope that you'll find new ways of finding joy *during* care, even in unexpected places, that that is worth exploring in your life.

Authors Rooted in Care and Their Lives as Caregivers

Both of us are versed in family care from our life experiences and professional backgrounds. Gael, author and founder of the Yokibics Mindbody Institute, has been a spokesperson for complementary health at the junction of medicine and self-care. Gael is a yoga therapist and Certified Caregiving Consultant, which means she coaches those with emphasis on caring for caregivers, and with her own background, she does it through the lens of her lifelong holistic self-care models. Gael feels passionate that sharing a set of principles surrounding family caregiving, which includes the caregiver in the center of the circle of care, can be less daunting and more fulfilling for a caregiver. Learning how to care for others is a path that has been walked by others who are reaching for your hand.

Sharon was called to write this book in part because of how importantly she takes the idea of training others on resilience—but also because she has a doctorate that focused upon family care of dependents who are older or younger. She also has extensive work experience in the corporate workplace as a trainer and project manager, and she has an interest in helping teach families and delivering training workshops for parents and teachers. Sharon has three kids of her own, and her son with experienced some delays in speech and sensory areas and had used the services of a local Regional Center who developed a plan of interventional service delivery until he reached age 3 and his delays caught up and he grew out of the need for services. Sharon's parents also both had Parkinson's disease, which caused disabling physical ability and struggles that included finding a residential care community for her elders and hiring care when the family wasn't able to handle the tasks the facilities couldn't. Sharon had to learn about caregiving after a traumatic brain injury from a family member's fall and ways in which to navigate from palliative to hospice and finally death and bereavement in the midst of a family member's life journey ending.

Gael was a Sandwich Generation Caregiver, a Baby Boomer who raised two children and buried two parents while being an independent entrepreneur, or solopreneur.

In November, 2000, Gael's twenty-three-year-old son was living across the country, successfully navigating an exciting new job. In the blink of an eye, he suffered a severe traumatic injury while engaged in his passion of skiing, and Gael was suddenly catapulted into a whole new life of urgency and care. Her experience with her son was both medical and institutional: life-support, trauma centers, multiple tiers of hospitalization, inpatient rehab, outpatient rehab, and finally the aftereffects. Today, Chris has an amazing recovery story as he carries on an exceptional life, yet as his parent, Gael is keenly aware of the journey. Subsequent to her return home, first one and then the other of her parents entered the long, slow process of eldercare, which required recognition, intervention, communication with siblings, and the eventual sale of their home with a move to assisted living. This kind of caregiving was a family affair, and engagement with hospice was the final choice her family made as they kept an intimate and close personal relationship with each parent. Her experience has been an ever-expanding study of what caregivers can do to support themselves and their loved ones within various medical levels of care, what to do when change is needed, and how to empower oneself to question when and where to let go or lean in.

PRAISE FOR THE AUTHORS

Praise for Sharon Zint Marts

I have shared the privilege of hearing Sharon Marts speak in a workshop about caregiving and can attest to her knowledge, sensitivity, and ability to get things done on projects. Sharon comes from a very special tradition of being cared for, caring for, and now teaching those who are caring for an elderly loved one or patient.
—Caroline Allen, director of social services at a senior living community

I admire the passion Sharon Marts has for assisting families in making informed decisions for their frail and declining loved ones, as well as caring for themselves (as many times they are the caregiver). Sharon imparts this information through her seminars, workshops, and personal experiences. Sharon provides information such as where to get support (group support) and hiring caregivers. It is so important to give family members information and resources so as to make informed decisions. This is a huge concern for many families, and Sharon is passionate about providing and imparting this information.
—Linda Zybach, RN continuing care nurse/case manager at a hospital

My wife, Sharon Marts, knows well what I see in the emergency room every day. She offers her education and expertise about family care to many in our community, which is invaluable for families in crisis who have medical experiences that are changing—often for the worse.
—Dr. Bruce Marts, emergency room physician

Sharon's expertise used to help a Parkinson's fund-raising committee brought a much-needed perspective: that of the caregiver. Her personal and professional experience in this area immediately expanded our outreach to those with the disease.
—Andy McSunas, Parkinson's disease fund-raiser committee leader

Sharon is among the strongly committed intergenerational caregivers, believing that assistance is a not just a policy area, but a moral decision rooted in the writings of philosophers and theologians and played out every day in families. At the same time, Sharon couples this moral dedication with a scholarly commitment to translational research on caregiving to advance and improve the outcomes of care recipients and caregivers.
—Dr. Susan Stryker, university professor of Entrepreneurship in the School of Management

Praise for Gael Chiarella Alba

Gael is a poet of mind, body, and life's experiences. You will feel immediately strengthened working with her. She is quick-witted, practical, clear, wise, and kind, and she wraps you in the feeling that your ideas and instincts are exactly right. She has full faith in you. Work with her—her words—and you will gain energy, peace, connection—wherever you are in life.
—Bob Soulliere, founder of Breathe Your Power (breatheyourpower.com), certified Wim Hof Method instructor, and CrossFit L1 trainer

As a caregiver for my mom, who had a long history with MS, and as a daughter who needed resources, Gael helped me nurture myself in order to do my job, both personally and professionally.
—Lorraine Aguilar, physical therapist and yoga therapist

Gael, I just wanted to thank you for posting about caregiving on Healthcare 3.0. As a Reiki 2 practitioner, as well as a CMA and surg tech, I am one of the few who straddle the fence and believe a combination of therapies, as long as it's beneficial to the patient and their recovery, is all that truly matters. Thank you again!
—Health 3.0 advocate

Gael, how I wish you coached me when I was caregiving! I love your calm and centered presence. Your advice is spot-on. I like your "lucky space" parking, how you encourage us to have a self-care routine, and that our choice is just that—our choice. In life, there is always joy. Thank you!
—Colleen Kavanaugh, host of The After Life podcast, certified dementia communications specialist, certified dementia practitioner

I'm living it and learning how to be better at it. It was really difficult in the beginning. Thank you, Gael, for the work you do to help caregivers who never planned on this role.
—Beverly Pipes, internet marketing and advertising specialist

CHAPTER 1
Multigenerational Care

I care for you, because you are mine, and I am yours.
—John McKnight, *The Careless Society:
Community and Its Counterfeits*

The Realization: Gael

My middle sister is speaking from the pulpit on the occasion of my father's death. The details are familiar—his childhood, his life, his work, and his contributions. I listen to the story of a timeless hero, and in so many ways, he was. Yet, as his eldest daughter, I find deeper reminders of his discontent rising up in me over and over again. My heart is still as I try to remember - when did that part of his journey begin?

I am eleven years old. We've moved from a six-story apartment building in Brooklyn to Queens, in what my parents call "the country," where roses bloom on our lawn and the apple orchard at the bottom of the street offers the mystery of a forest. My younger siblings, six and three, are playing loudly as Mom, pregnant with child number four, is working to get dinner on the table. Meal preparation is hardly her strong suit, even on a good day. Dad arrives home via bus and train from Manhattan, looking dashing in his Mad Men hat, suit, and tie. He has plenty to share about his day. Plenty. He needs an audience, and as I witness my mother's distraction, I realize I am becoming that.

Now I am twelve, thirteen, fourteen years old, and Dad, still growing as an executive, is increasingly dissatisfied. Things for him are changing, as his disgruntled commentaries attest. He arrives home from work—often late—and, as always, filled with the need to discharge. By now, it has become more a compulsion to set "those people" straight. Mom, with a baby now, is unable to listen properly as he follows her round and round the kitchen—talking, talking, talking—and then takes a seat in the middle of it all, talking some more.

The kids remain mostly unseen. Mom is overwhelmed, and I encourage Dad to step into the living room. "Tell me," I say, and he does, as I listen instead of sleep. My inner caregiver is already being shaped.

My sister has grown big enough for outings, and I excitedly collect her after school, going right back out to push her in the carriage around the neighborhood. She is privy to all my schoolyard secrets. We become so attached that when sitters arrive for parents' night out, she awakens and cries out for me. Just me. "Why do we need the sitter here at all?" I grumble, not understanding the grateful payment my parents' dish out when it is I who has been tending to the care of the little one.

Seeds of awareness begin to emerge for a call to balance the scale that has already begun to tilt with a need to take care of myself. I do learn. I learn to create my own places, both inside and out, where I am free to listen to myself in quiet, or commune with voices of optimism and light. I practice my skills. I practice them a lot, and I find that the demand for this sort of care can be daunting. Worse yet, taking good care of oneself—body, mind, and spirit—is often a maligned priority that is downright misunderstood.

Then again, learning the difference between selfish and self-care can take shape surprisingly well as the foundation of an entire career, and from that, a life well lived.

Care Crossing Generations

As in Gael's story, care crosses generations. Each family caregiver's life may be constructed differently, but they all involve looking after the needs of others. Caregiving gets its credence when the care of another is challenging from medical needs or complexity. No matter what our places in the familial line, we are caregivers when the majority of tasks fall into our purview. Through our care, we are shepherds of the health of another—not so much the progression of a condition or disease but the attitude surrounding the journey. Through our caregiving roles, we can imbue an outlook that shapes whether the care of others is done in a way that is rote and methodical or meaningful and hopeful.

Most crucial of all tasks as a family caregiver managing intergenerational care is knowing how to navigate the systems, medical environments, and circumstances of the person each of us cares for. This demands managing new choices and being aware that the family caregiver is the center of the care circle in every step of the way. These tasks can be comprised of transportation, appointments, accommodations for social engagements, and even setting up legal and other permissions (i.e., power of attorney) to facilitate decisions.

Every potential caregiver needs to have a framework of basics to ensure the successful delivery of the care that will be required while maintaining his or her own footing. In this chapter, we explain how to manage the complexity of the situation. You'll see how care of even one person affects the care of all in the family.

If you don't already see yourself as central to the care team, this chapter allows you to see why that positioning is critical. Finally, you'll begin to see how gaining caregiving skills can save time. It may mean doing research so you don't have to approach a doctor or specialist or taking a few steps to set up an easy-to-maintain planner for carrying essential information and papers. Know-how can ease suffering and reduce the stress of all concerned. But first, it's vital to see how the complex care of a family member affects everyone in the family.

Family Care Affects All in the Family

This chapter focuses on intergenerational care and is vital to explaining how caring for older, frail family members in different generations influence the *whole* family and not just the immediate family giving the care. Dr. Ira Byock, author of *The Best Care Possible* (2012), is a physician and at the highest level of management at the Institute for Human Caring of Providence Health and Services. Every day, he sees how the caregiver and the whole family are affected when a family member has a chronic illness, disease, or condition: "When any one person becomes sick, a family inevitably experiences the illness." His words reflect how families are interconnected and how a decline of health can take a toll on all in a family in ways that are sometimes not felt at a conscious level.

The term, *Sandwich Generation Caregiving,* initially used by Dorothy Miller in 1981, refers to care of both younger and older generations and is one type of care crossing generations. *Compound Caregiving,* as reported in a study of this sort by Perkins and Haley in 2010, is another type of care that describes when those who are lifelong caregivers already, such as with a special needs child, then take on additional challenges of addressing needs of another family member. These are some examples of care with implications that can ripple across a family.

Family Caregiver's Role Among the Care Team

This bears repeating throughout the book: the role of the caregiver is a central one. Aside from the patient whose needs are being met, you might think the center of the care team driving actions is the doctor, the hospital, the nurse, the machine, or the medication. Not so. They are the tools, the distribution channels, the systems, and the methods. They are the deliverers and the how-tos of care. The caregiver is at the center of all of it, alongside the patient whose needs are demanding attention. The role of primary caregiver is often assigned to a parent, spouse, adult child, or nonfamily member who has power of attorney over the patient's health-care decisions. This means being more than transportation, companion, or witness. Additionally there can be non-primary caregivers within the family circle. The family caregiver is an engaged participant who needs to know what is involved in a caregiving role and who comprises the care team—the group of people who carry out the care of the family member in need.

Because many caregivers don't recognize their importance, they may not perform crucial functions to support a family member whose health matters are becoming more and more complex over an indefinite period of time. Caregiving expert Colleen Kavanaugh is a certified dementia communications specialist who knows well the skills caregivers need. She states that educating our caregivers on how to be informed advocates is vital. "The role of the caregiver is overlooked, and if we can urge family caregivers to clearly communicate how they would like to create a team with the doctor / medical staff, some of the frustrations in not having a previously unspoken wish acted on can be avoided."

Avoiding frustrations and mistakes can be achieved by driving the direction and knowing the path forward, which begins with compiling information, asking many questions, and being a vocal and engaged communicator.

Communication Is Key

To begin to evaluate areas in which communication is vital, look at those with whom you interact on a day-to-day basis to talk about your family member's care. Think about a typical day and also one in which harder tasks come into play. Also think about a day you take off, when someone else is handling care. Take into consideration the big and small functions that make a difference between something being smooth and challenging if the steps are not in place.

As you read this book, you will learn to improve the skills relevant to your role as a caregiver and see a difference in how you deal with family members and medical personnel, hire paid care, and even discretely convey your caregiver experiences to those in your workplace.

Your Approach in a Time of Caregiving

As you interact with others, don't forget that kindness and respect are often the most powerful gifts you bring. Your attitude and actions may be fraught with frustrations and impatience, but kindness is a buffer that can aid all in challenging times.

Keep in mind to use your own intentions for clarity and peace when relating with:

- family members whose care directs this journey
- paid caregivers and service providers
- other patients
- most importantly, yourself

Broadening the Perspective of Family Care

The existing health-care system is changing quickly. Often methods are focused upon cures and treatments, and many have hard choices that require people with care needs and their caregivers to be ready to commit and question. For those being treated, it may not be obvious that more than one way forward is a choice. Recent books by medical professionals attest to this. Atul Gawande's *Being Mortal* (2014) and Paul Kalanithi's *When Breath Becomes Air* (2016) talk about the challenges of quality versus quantity of time left in lives cut short by cancer and how best to proceed. These books offer the complex decision-making that may seem clear-cut, yet from these doctors' own admissions, are often gray areas with room for families to disagree or change the path forward.

A book about care and the eventual death of her mother by poet and author Meghan O'Rourke, *The Long Goodbye* (2012), offers a powerful account of life outside of the doctor's office in a home where this poet who is also a daughter finds meaning from the experiences in care for her mother.

Rethinking Caregiving through the Lens of Meaning

It goes without saying that care of family members of different generations can be hard and not always rewarding. Yet without taking the time to examine notions of care, this time of caregiving for one or more generations may lack value in one's eyes. An exploration of meaning offers new ways of thinking, and ideas of that impart resilience about a different hopeful future.

The reflective questions posed in this book are intended to help you explore those meanings and find that renewed resilience in the care of others and particularly in your care for yourself. There are fruits to this care, which you might not readily see at this moment, but you will if you tend to the deeper aspects of what matters to you and your family members with intention, an ethical understanding, and a recognition that this time of care is valued so much more than you know.

Our entire health-care system could not run without you. You are a force to be reckoned with, and you deserve the care and support of a society that relies on your care. Know how very important you are. Perhaps this time of caring may be seen not only for the duties of care given to others but also in recognition of how vital our simple presence with others can be. Through everyday actions with an inward focus, you will find a renewed impetus to better recognize yourself as a caregiver while supporting other family members who require care.

Chapter 1 Reflections: Multigenerational Care

- What familial memories from the past and hopes for the future inform your understanding of family caregiving?
- How are those memories and hopes informing your choices today as a caregiver?
- How are the act of caregiving and your role seen by society? How do familial memories inform your sense of what is the right thing to do for a family member needing your care now?
- What do you need to do or have in order to effectively communicate about your caregiving situation?

Notes to Myself

CHAPTER 2
The Caregiver

We hold life and death in such separation, and so much about the
connection between illness and caregiving is kept out of sight. I
feel the call getting closer to find ways of bridging the gap. I want
to invite us all to see, to prepare, and maybe even, to practice.
—Gael Chiarella Alba

Separation: Gael

*I am curled up reading in one of the family lounges at the prestigious
rehab hospital for severe traumatic injury halfway across the country
from my home. My son has made it through the posttraumatic
resuscitative heroics that occurred during the days following the
accident and settled into the calm but lengthy road to repair.*

*In truth, after weeks of this, I am here for myself as much as for
him, for his trauma is also my own, and there is simply nowhere else I
can be that brings me comfort. I have habituated to hospital life as best
one can in circumstances such as these. My business has come to an
abrupt halt, on pause, as is my mail, and I am suspended in the needs*

of the present moment. I know this will be so for some undetermined amount of time.

I glance at the bulletin board, which is filled with announcements. Educational workshops are offered for patients and family members, and I am in the habit of attending every kind of course that's offered whether it pertains to us or not. Sometimes it is just for interest's sake. More often, I am adding yet another layer to the arsenal I am building for the potential of future use.

Today's topic is "How to Manage Bowels When Wheelchair Bound."

One thing about hanging around a place like this—you are introduced to topics you never thought existed, and opportunities to learn are anything but ordinary. Today, I am the first to arrive, taking a seat against the wall as one by one, mostly young male patients file in, jockeying wheelchairs into position in the rather cramped room. No one seems to know each other, and surprisingly, no one has any supportive company with them. I have learned that this institute is filled with "risk-takers"—vital and young, and I am again sensing the need for increased support for patient advocacy, something I experience ten times a day.

I am struck by an overwhelming gratitude that my son, who has continuously and deftly moved along in his own recovery, will have no need for this training, yet for the dozen or so young men in front of me for whom this is not the case, I am here to witness their moment and learn.

The nurse begins her introductions, and looking new at this, calls the group to attention. "Who among you has ever sat in a wheelchair?" Warming up, she is looking for a place to start. Everyone looks blank.

Could this really be? I am flashing to my grandfather and my own familiarity of long ago helping get him out of his chair and into the bed with the fun of taking complete charge of his vacated chariot as I maneuver it this way and that through the squeeze of a three-room apartment.

"No one then?" the nurse asks. She is recalibrating. "All right. Who has pushed a wheelchair for someone else?"

Memories of the old neighborhood, the store, the doctor, the many hospital hallways collide in me. Lots of pushing for a young person to do.

"No one? No one at all?" This nurse is striking out. "Well then, how many of you know of someone, anyone at all, who has used a wheelchair?"

More silence until, finally, a hesitant voice breaks through. "Once I saw an old guy down the aisle at the drugstore. I didn't know him, but there he was, just sitting there in one."

So begins the meet and greet for a training that will initiate each of these young men into skills they will need for the rest of their days, lived in chairs they know nothing about, as we all quietly settle into an unmistakable awareness of just how big the gap really is between what is familiar to us today and what the necessity actually is.

What is a Family Caregiver?

What is a family caregiver? It is worth describing what this care is since many see only *paid* caregivers in this way. Most typically, a caregiver is a family member who is unpaid (a spouse, partner, family member, friend, or neighbor) and is involved in assisting others with activities of daily living and/or medical tasks. It is possible for this to be confused with hired caregivers. This designation is not a label that is singular. It pairs with the role of parent, spouse, adult child. Not claiming the role of caregiver may mean that you are missing part of this identity, which involves being a part of a bigger demographic.

A 2015 report by Pew's Research titled, "5 Facts About Family Caregivers" was authored by Renee Stepler and indicated that there are 40.4 million unpaid caregivers in the United States, and nine in ten are providing care for an aging relative, namely caring for a parent. This number has only increased to be closer to 44 million in 2019. Even more family caregivers take care of a non-elder adult like a spouse or adult child, and some care for a child living with a disability or long-term health issue. Although many American adults do not currently provide care to a loved one, that can change over time with a condition or chronic illness or accident changing the course of events in a family.

A 2018 report published by AARP specifically concerns young adult caregivers and is titled "Millennials: The Emerging Generation of Family Caregivers." Using data from the AARP Public Policy Institute's National Alliance for Caregiving in the United States (2015), the report notes that about 76 percent of the people cared for by millennial family members are fifty or older, and the average care recipient is sixty years old. The average care recipient helped by a grandchild is seventy-seven years old. And more than half of millennial family caregivers (51 percent) are the sole caregiver, alone in their duties.

While the older generations are thought to be taking on most of the burden, this report shows how the caregiving role is shifting to younger generations with ages eighteen to twenty-four making up 35 percent of caregivers and twenty-five to twenty-nine comprising more than 31 percent of caregivers. Further demographic breakdown is shown at www.aarp.org (with search term: millennial family caregivers).

When Family Care Goes Above and Beyond

What defines care that goes above and beyond what we think of normal expectations? Is it checking in on grandparents or parents who are elderly? It is when the care involves challenging medical prognosis, weakening health, and behavioral and educational complexities that require supervision by a family member? This kind of care can also involve special needs care—whose medical needs are vastly different from raising "typical" kids without special needs.

Regardless of care time, for many caring for a person with a chronic or debilitating condition, this care is never static, and it can involve a fragility that can be altered from moment to moment. When a medical condition is affecting a family member's quality of life, the reliance upon one another shifts the balance in the relationships and the family.

Caring for a Newborn versus an Elder with Special Needs

Many times, a parent gives birth only to care for a family member for the first time. Some new parents find their tiny baby's life will unfold differently than was expected with news of a special needs diagnosis. This is a different type of family caregiving because the term "new parent" fits, but it is like other forms of caregiving in that the parents are wondering if their baby would one day exit the hospital and have lifelong health problems or developmental delays that will affect the short and long term and mean a life of disability.

Parents of babies born into the world with challenges are thrown into a similar caregiver role as others who care for different ages. Their life becomes dedicated to their family member's health and comfort as this care becomes like a job. Just like a family caregiver addressing the needs of an elder or spouse or aunt who in adult age has health conditions needing attention, the caregiver can be found making calls to specialists, interfacing with insurance companies, and keeping detailed health records. Yet with the growing infant making its way in the world, the parent/caregiver also works for educational options that are guaranteed and yet often not available without a fight.

Caregiving: Choice or Forced Responsibility

Whether you agree to be a family caregiver or find it thrust upon you, know that you are not alone. Most all working adults will at some time become caregivers for elders. Numerous families with young children face developmental delays that may or may not turn into disabilities that are physical, cognitive, or show themselves in the classroom as a learning disability. Car and other accidents will cause short-term caregiving issues for some as they heal or require long-term stays in recuperative post-care facilities and a new life with a disability after. There may be those who *were* caregivers who find themselves being cared for. Whatever way in which we attained our caregiving role, this is for sure—caregiving will affect relationships with all family members and will require different skills than were needed before taking care of a dependent family member.

Many family members elect to be a primary caregiver whether volunteering for the role, through proximity to the family member, or as a devotion to helping a family member in need. In some families, unseen but intractable factors are at play such as cultural expectations about the role of care, including which sibling should care for family members in need. Whatever drives our decisions to care for family, there will be changes during that journey that may alter one's view of care. There might be initial deference and passivity in the relationship or short-lived harmony. Assertiveness becomes a necessity since the caregiver role requires advocacy and decision-making skills that cross the lines of people, financial decisions, and medically ambiguous choices. The caregiver can feel pressure and stress that will increase if the caregiver isn't aware and ready to respond proactively.

Family Care and Outings Beyond the Home

Errands and leisure activities can be inclusive. Caregivers need not be boxed into life that is limited, but can discover places to go outside of the residence. With a little research in advance about planning an outing in light of disability access, more possibilities become apparent. It will become easier and easier to request accommodations as venues and locations to visit will begin to expand. This doesn't preclude a required sense of fearlessness!

In Sharon's eldercare circumstance, both parents used walkers and wheelchairs yet went out to dinner often, bowled at the bowling alley using assistive equipment, and even went to the beach (beach wheelchairs work beautifully on sand) and to the movie theater. Often an outing involved finding an inclusive and accommodating location, and in most cases, it was surprisingly easy to accomplish! Although Sharon's parents' Parkinson's symptoms included tremors, saliva dribble, and the need for frequent bathroom breaks, she became adept at managing trips with an intrepid traveler's instincts. There were many instances in which people stared, but there were others when people cheered. Regarding one happy memory, Sharon's says, "I was in a race with my dad with him in his wheelchair and me pushing him in a one-mile fun run at a military base on Veteran's Day. He was high-fived and applauded, and at one point, an active-duty military official asked to push him in my place. That moment of joy reinforced the worth of trying to get out and do more things that included my disabled family members." Trying to create experiences that involve going out may involve challenges but these can be addressed, and sometimes with serendipitous circumstances.

Sometimes accommodating and inclusive places are not evident initially, but with a commitment to seeking out these places, the effort will pay off generously. Keep in mind that strangers who are unfamiliar with disability in their lives may unconsciously be fearful of someone who is different than themselves. This may become evident in awkward and nervous behavior but this can be overcome when people are exposed to inclusive situations and their familiarity increases. This disconnect is particularly awful because caregivers already have so much to balance and manage as it is, and the rejection of others (even with only their eyes) feels like something to avoid. But don't let this stop you from trying. Persevere by finding examples of others who are succeeding.

Sadly, the way we treat the most vulnerable members says a lot about our society. It likely won't change overnight, but that shouldn't be an impediment to trying for inclusion. Sharon's *My Grandpa Wants to Get a Porcupine* (2015) addresses inclusion and how young kids can come to view disability through a different lens than one of fear. Sharon's own children were fearful of their grandparents who had a disability, and writing the book was a helpful reminder to her kids and others of how inclusion begins at home and goes outward from there.

Fearless caregivers exist and are examples to learn from. Parents of kids with autism often learn about how to work with their children to manage behaviors or carefully select environments where behaviors won't be an issue. There are events and programs that foster inclusion by welcoming families with medical challenges. Although the age groups of parents with special needs kids and adult child caregivers caring for elders may not interact, their end goal of having a family life that allows for inclusion and family togetherness are the same, and both groups benefit from learning about accommodations.

The Journey of Family Care

When life revolves around upholding another's health and well-being, you may find yourself on a difficult path that diverges from your peers. It is also special in a way others don't understand. You come to know your family member and take part in their life journey in a different way. You love them "with a different kind of fierceness," said one caregiver.

Caregiving actions are intentional and rooted in the greater good of helping our family members. We can't forget to help ourselves on this journey too. Whether caregivers are "fierce" or not, their actions go beyond the completion of tasks in the context of a family. It is care that is personal and meaningful above all.

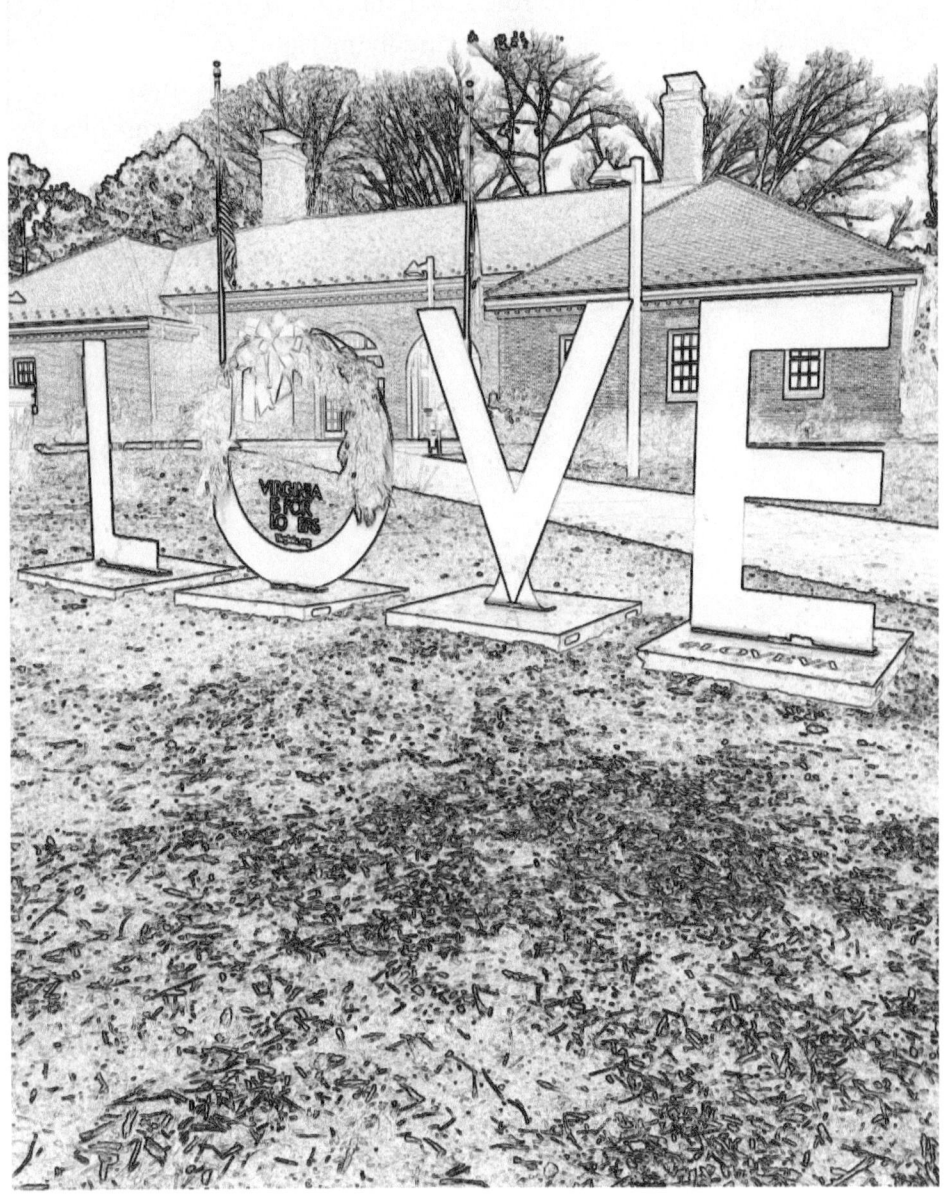

Chapter 2 Reflections: The Caregiver

- If you let go of expectations and "shoulds of perfection," what might your caregiver role look like?
- What is the "work" of your caregiving right now?
- Suppose you have a friend who has suddenly become a family caregiver, what advice would you give from your own learnings?
- What description best suits the role you have now? What title do you think would be most empowering for you?

Notes to Myself

CHAPTER 3
The Value of Our Care

An inconvenience is an opportunity wrongly considered.
—G. K. Chesterton

The New Bed: Gael

Today my son has been chosen to be the lucky patient who will test-drive the fancy new model bed the hospital is considering. It's one of a kind at the moment, and it has all the bells, whistles, and high-tech gear one could hope for attached right to the frame. The selling points are low maintenance, state of the art, all-inclusive, all good.

This is an exciting day. Chris has gotten progressively stronger, moving from one unit to another—one floor to another—one room to another, and today, we are going downstairs in the new bed. A fleet of six or so personnel come in to support the move, some detaching life-support machinery in return for a nurse with a hand pump, while others rearrange poles and tubes, and briskly wheel him out right on his enormous bed.

Several staff hover by his side as we press the elevator button awaiting the whoosh of its arrival. The door opens, and ready to go, the entire team realizes simultaneously that the bed does not fit in the elevator.

THE BED DOES NOT FIT!

Stuck in the middle of the hallway, the strong nurse is hand-pumping oxygen to my son, and both are sweating. A quick-thinking crew member races down the hallway to the freight elevator, calling for help as he goes to empty stuff that seems to be stored there. Frantically, and finally, it is freed up so the bed, tubes, nurse, pump, crew, and son can thankfully—and blessedly—get shoved in.

New beds and old elevators, when bigger and better is not that, and all that glitters isn't gold.

We did eventually make it to our destination successfully, but it was not without struggle. My son was reconnected to the machinery and automation of buzzers, bells, and breathing in his new room, but the whole affair was, in truth, another sort of trauma for all of us. Even the best-laid plans can go awry, and the solutions that show up demand resilience or even a brand-new approach.

Recognition? Who Needs Recognition?

If life is already complicated with care needs in a family, then the idea of recognition may seem like an unneeded pat on the back, but it is much more. If a worker at a company worked at the pace of a caregiver in the height of troubling times, they might be eligible for an award! Think through what a caregiver does:

- You are a caregiver through thick and thin—while rarely finding a replacement.
- You are devoted and loyal, showing up to do the best you can for your family member.
- You taught yourself how to provide care in an overwhelming health-care system.

- You learned on the spot, swallowing the stress because you had to.
- You literally make another person's life better with a focus upon their health (often by sacrificing yours).

Recognition is a hard word because of the associations related to earned recognition or assigned value and praise, which we don't normally associate with care of family. Perhaps a better way to explain this concept is by comparing it to visibility. To be seen is to be understood. In *The Invisible Caregiver: How Can Doctors Care for Them?* (2001), Delia O'Hara stated,

> Because unpaid caregivers—men and women, relatives ... of the grievously ill—are not counted in the health-care ranks, they might as well be invisible. Their needs are seldom considered. Frequently nobody even bothers to show them how to do the tasks they have taken on out of love and a sense of responsibility—often at a huge cost in terms of time, money ... and their own interests and health.

To be seen in this sense is to be counted in a visible way, and caregivers can't be known unless they are formally acknowledged. Thus, in a philosophical way, to recognize a caregiver is to make visible at a higher level, which assumes a worth having an intrinsic value. This value is hard to quantify societally, but in most families, caregivers are invaluable.

Caregivers Are Invaluable

In 2015, American Association of Retired People's Public Policy Institute authored a study called "Caregiving in the United States." In the report, they estimated that 43.5 million adults have provided unpaid family care in the past twelve months—many of whom don't even consider themselves to be "caregivers" by title. This number has only grown. On average, those in the AARP study indicated they spend 24.4 hours per week caring for their loved ones, and their tasks range from small ones addressing transportation and housework and complex ones handling tasks paid nurses usually perform, including helping with injections, feeding tubes, catheters, and colostomy care—all unpaid and untracked by the systems our government uses to gauge problems based upon the money they cost to fix or replace. Caregivers in families have a role that is largely unseen, and accordingly, they often feel invisible and unrecognized.

Seeing Ourselves as Caregivers

Being seen by others as important is one thing, but sometimes caregivers don't see themselves with the recognition they deserve. Taking the time to understand why caregivers are vital in the system of our loved ones' family care is important. Medical professionals look to the caregiver for direction and authorization. If caregivers see their role as having authority, they may question something that would have been disregarded in prior situations.

Caregivers can unknowingly lessen their own worth if they can't find their voice. Perhaps it is the time that has elapsed, or the drudgery of the tasks, or surrender to the invisibility of the situation. Sharon speaks to support groups, and people often explain that as they sit in the back of the support group meeting—where the emphasis is focused solely upon their loved one's condition—they feel like they matter little. Caregivers need to be taught to advocate and speak up and be given permission to own the task at hand. Caregivers talk about how hard it is to ask for help from other family members or how they face anguish and guilt when they do find the courage to hire outside care.

Taking the time to own the title of caregiver and the role involved speaks to the topic of permission to do more, and it involves doing less when it is appropriate (without guilt). Taking care of yourself, also known as *self-care*, is an important piece of self-recognition. A small action as taking time for one's self for quiet, saying no, or attending a workshop or support group intended solely for caregivers can make a difference in the way caregivers see themselves and recharge their energy levels.

Caregiver Burnout and Its Effects

Sharon studied caregivers in 2013 in a University of San Francisco doctoral dissertation published by Scholar's Press. In *Bridging Generations: Personal and Societal Implications from a Study of Sandwich Generation Caregivers*, participants shared stories of their lives during times of caregiving stress and how they often struggled to recognize the cost. A participant named Lisa explained that she sometimes felt, "overextended and pulled as to who needs more attention at that moment … [thus it] makes you feel guilty that you can't do it all. So I think that is when you need to ask for help and think you need to be self-aware because of caregiver burnout … And then if you burn out, you're not going to be useful to anyone."

Another participant, Paula, knew caregiver stress well. She mentions how exhaustion from caring for family members in need led to her own hospitalization, and she was not able to help herself or others. Story after story involved caregivers taking on a toll they physically and emotionally could hardly shoulder alone. Some got help, and others muddled through—sometimes harboring resentment and frustrations that lasted beyond the time of care.

Caregivers are human, and too much stress and exhaustion take their toll unless caregivers know how to stand up for themselves and their own well-being. Knowing how to be self-aware and advocating for yourself as a caregiver is vital to being both strong for yourself and others. Paula added, "I take each item as it comes, never forgetting to stand up for myself along the way. If I don't say anything, people won't know [that I am burned out]. I always try to be understanding to others and try to teach others to be understanding to me as well."

Only with a sense of self-recognition as caregivers can we honor what our needs are so we can remain strong and dedicated to our families and not lose ourselves in the process.

The Give and Take of Caregiving

Recognizing the value of our own care and having society see and value it are two different things. When we are able to see these invisible caregivers in terms of the benefits they give society by taking care of their own, a different recognition might be ascribed to these family caregivers. Awareness is a key part of showing those in society how family caregivers contribute through their valuable role. Perhaps training and education, government or private services, and workplace policy changes may one day increase the exposure about what these caregivers endure. These matters will become more important as more elders age and more people in society experience this predicament. If a caregiver lacks support or the time to find it, perhaps if they look inward to find a deeper and more meaningful understanding of the family caregiving experience, they may see their role differently, better recognize their role, and find renewed capability.

Delegating Care

While acting within a family with no help nearby, delegating care might help the family and reduce one's own burden. Perhaps a caregiver might meet his or her own personal needs by hiring care in order to take a break. This kind of self-care after a prolonged period of caregiving helps the self and shows others how the role is not limited to family. Hiring a caregiver might ease the load of the one who has no immediate help while offering them a chance to attend a support group, experience commiseration among caregivers in similar circumstances, and offer ways to count on others to lessen the toll of care.

Chapter 3 Reflections: The Value of Our Care

- Having written down your hopes in the prior chapter, do you feel you have aspects of care you can or cannot offload?
- How would you view your care if you delegated what you do to someone else? How would the care change—for worse or better? Why or why not?
- What does it mean to you to find your voice?
- What is most significant thing you'd like to see for yourself in order to validate your worth as a caregiver? What else?

Notes to Myself

CHAPTER 4
Something's Gotta Give

The moment you realize you just spent the last thirty
minutes arguing with your loved one about something
small, only to realize your own dress is on backwards.
—Sharon Marts

Losing My Bearings in the Midst of Care: Sharon

*There was a time when three kids' care was the center of my life.
Parenting through ups and downs was hard enough, or so I thought.
One day, I got a call at work from my parents' friend in Colorado: "Your
parents can no longer live safely at home. You need to do something
and soon."*

*After many weeks of convincing my parents to move, we began
transition planning with my parents and siblings. Amazingly, within
six weeks, the day came for me to move my parents from Colorado to
California. My mother was nauseous and dry-heaved throughout our
car ride and plane ride from nervousness. Both of my parents needed
wheelchairs, which made me wish I had four arms to push them myself.*

Instead, I relied upon transport helpers at the airport for our six o'clock in the morning flight from Colorado to the plane and then from San Francisco's airport to the parking lot. I burst into tears when we made it to the car. I got my parents settled in and then took a minute outside of the car to cry alone where they couldn't see me.

This was the beginning of the caregiving for my two parents who both had end-stage Parkinson's Disease—and what a wild ride the next four years would be. Thankfully I did not have to do it alone and had the support of my three brothers and numerous friends and helpers (paid and unpaid).

As my parents settled in to what was the first of three facilities where they would reside over four years, something shifted within me. Not even a month after their big move, a sense of anxiety became permanent within me. Mentally, I was anxious and found myself having jitters or stomachaches when worry would creep up. I manifested my worry even in my dream state and would sometimes sit erectly in my slumber and give off a bloodcurdling scream—without awareness that I had done so until my husband told me about it the next morning. Sleepwalking was another sign of my stress. Once I walked to the end of my staircase (asleep), only to fall down nearly half of the staircase, and managed to catch myself and come back to bed. I had no recollection of my sleepwalking and fall until my teen daughter recounted to me the next morning what happened (and this explained my bruises). My nocturnal sleeping issues were only a small representation of my stress during the day. The toll on me was great, and my brain lacked the ability to turn off properly. This went on for two years before I got help for myself with a doctor's visit focused upon me as well as meaningful steps toward self-care.

How did I cope when I did get help to regain my bearings? I joined a monthly caregiver-support group. The people I met there helped me to re-find my sense of my old self which seemed to be getting away. With other caregivers as my ear, I felt I could speak candidly. We became close enough that we could talk often and not wait for our meetings. I learned from my caregiver group to let go of commitments that drained me, to seek out things that gave me energy, and I even hired Gael as my

Caregiving Consultant so she could coach me through a hard time. I came to see from others in my caregiver support group that asking for help was not a sign of failure. I also made appointments with my health plan's counseling department, and I worked with a hired geriatric care management to help direct my parents' care so I could remove myself when I needed to. I also emailed and called my siblings to vent, ask questions, and delegate. As long-distance caregivers, my three brothers were amazing and they never failed to come to my assistance every time I asked them to (but I had to ask). Over time, I regained a sense of control.

Are We Prepared?

Caregiving is stressful. It's not uncommon for caregivers to be more in tune with the person they care for than for themselves. Signs that your own health is starting to falter may not even be obvious to you. Things that may be different and may or may not appear noticeable to you include social withdrawal or not enjoying events you once regularly found pleasurable. You may also have wishful thinking or find yourself evading negative thoughts and emotions.

Denial, avoidance, and disengagement are part of what is called "regressive coping" as defined by in *The Comprehensive Handbook of Personality and Psychopathology* (2006) edited by Jay C. Thomas and Daniel L. Segal. This type of coping is caused by stressful factors including those faced by many caregivers who find themselves worn down and handling their situations by avoiding or denying the circumstances. However, withdrawing from the reality tends to make things worse or push the difficulties down the road.

Caregiver Stress

There is a phrase known to some in the healthcare field as, "caregiver syndrome," or "caregiver stress syndrome" but this description has not been adopted formally into medical manuals. This characterization is a condition of exhaustion, frustration, anger, rage, and/or guilt that results from unrelieved caregiving –especially when caring for a chronically ill dependent over a period of time. This term is sometimes used by health-care professionals, but it is not well known. The Family Caregiver's Alliance reports that the following areas are at risk when family members are overwhelmed by care responsibilities:

- sleep deprivation
- poor eating habits
- failure to exercise
- failure to stay in bed when ill
- postponement of or failure to make medical appointments for themselves

The fact that family caregivers fall into the same patterns of behavior listed above is not a coincidence. The care of a family member happens in our homes and is unavoidable. Yet the response can vary, and not all circumstances result in stress if protective factors are in place like having friends to help or stepping back from a situation. How caregivers respond to stress tends to cause their outlook to change. Experts identified two types of coping that are common responses and are often indicative of the way caregivers react and respond when caring for family.

Regressive Coping

The Comprehensive Handbook of Personality and Psychopathology (2006) edited by Jay C. Thomas and Daniel L. Segal explained that a negative way that many respond to stress is called "regressive" coping. These coping methods often focus on the problem or problems at hand. Consistent and recurrent thinking in a circular manner (also known as ruminating) is a typical regressive response. This can mean self-pity, wallowing, or feeling stuck—none of which are useful or helpful to caregivers. What is worse is that this type of coping strategy may be cause a person to veer toward depression. This is not to say that hearing bad news and feeling frustrated are "regressive" by any means. Instead, regressive coping is an attempt at finding a solution which is thwarted or stalled. But know that more than one coping strategy exists! There is hope, but it requires effort.

It's vital to sense a change in yourself and monitor your stress level. If you are experiencing something different in your thinking, it may change the way you feel and act toward yourself and others. By acknowledging this shift in yourself, you can begin to make changes in your coping, improve your well-being, and more conscientiously support the person whose care you manage.

The opposite of regressive coping is transformational coping, which is also noted in *The Comprehensive Handbook of Personality and Psychopathology*. This kind of coping is characterized by focused problem solving. This means brainstorming solutions and having faith that multiple possibilities exist. When a person sees or even just senses that more options are available, it opens up new ways to adapt. When seen in terms of the family member's care, it may foster a sense of hopefulness to return. It's easier to engage with problems at hand with this type of coping strategy, but it requires work on your part.

Transformational coping involves reaching outward, keeping those social outlets in use, and maintaining perspective. When problems are kept in perspective and not always ruminated upon, different aspects can be seen. When problems are engaged in this way, they can be grappled with, and pros and cons can be thought through. More decisive actions may result from this type of transformational coping in which the circumstances are felt deeply and explored fully, which can't be done with the avoidance and denial marked by regressive coping.

Conditions for Transformational Coping

You must create space to cope in a positive way. Make time for friendships—even if you have to carve out times in the morning, at mealtimes, or on weekends. Watch your breathing to see when stressful situations cause you to hold your breath or take shallow breaths. Step back and think from a different perspective to avoid regressive thinking patterns. When you feel unsteady, you need to regain a balanced, healthy perspective to ensure that you are renewed and ready to continue to be a caregiver.

It's hard to break old patterns if you are accustomed to avoidance or denial, but first, you must face the emotions that make changes hard. What is the fear about? Do you have trouble asking for what you need? Journaling or talking to a friend or a professional can make this easier.

Your subconscious, also known as your "other-than-conscious mind," is responsible for the automatic control of important bodily functions like your heartbeat, your breathing, your metabolism, and your immune system. If you imagine your subconscious mind is like a computer, it can get clogged up with "bad programs" over time after constantly being bombarded by information, negativity, noise, and stress in our busy modern lives. Your conscious mind has a much smaller processing power than your subconscious mind, and it prefers to do one task at a time. However, it is not uncommon for most of us to multitask throughout the day, which creates more stress for the mind.

Meditation, Exercise, and Other Stress-Reducing Habits

For many caregivers who have the commitment and ability to take up the habit of meditation, it can help give the subconscious and conscious mind a rest. Tai chi and yoga help quiet the mind but may not be for everyone. Many who take up a practice find it offers an unshakable resistance to stress, but it is only one avenue to stress relief. Other options are burning calories in an exercise class with others or a walk alone to get away and get exercise. Making time and sticking with a habit are what is necessary—not so much *what* is the exercise selected.

A Shared Reality of Caregivers with Varying Durations and Intensity

Remember that caregivers have in common their shared responsibility of caring for loved ones, but their experiences can differ. Some caregivers may have to face an imminent problem that is navigated at first covering days to months, but then is reduced in intensity over the years to concern and readiness to step in again from afar as needed (but the medical matter may not be resolved). Sharon's son had developmental delays in his toddler years and went through evaluations and speech therapies that helped with communication and anger and impulse control. Behavior was always an issue in public situations, and the judgment of onlookers was always an issue, but Sharon's son caught up and his behavior normalized by age seven.

Gael and Sharon were caregivers for family members who had brain injuries from trauma and know about the initial period in which brain function is being evaluated closely. After stability is restored—and a different type of physical and occupational therapy occurs—a new normal for the family member is found.

Other caregivers have a prolonged long, slow road of chronic conditions that worsen or, as with elders, involve growing frailty of a parent that ends with death. Gael's dad experienced twenty years of slow, degenerative disease with heart surgeries and kidney failure requiring numerous hospitalizations, dialysis, and long-term caregiving. Gael's mother had a much shorter disease vector, remaining vital until the last months, ending with hospice care.

Chapter 4 Reflections: Something's Gotta Give

- Remind yourself of times when someone helped you through a long haul or when you witnessed it in your life. Caring for others is seen in the United States as a burden, but other cultures don't often see it that way. Perhaps rethinking your own personal concepts of care can help you see anew why you may resent care when it is possible to be seen through a different lens of meaning.
- Reshape for yourself how you can continue to care and still have a full life, which may mean researching help or delegating to others. This may be hard to manage, but care for others should not stop the life you foresee for yourself. These are not mutually exclusive, but you must plan, be patient, and make compromises. Do not give up areas of your life entirely since this can cause resentment and martyrdom. Allow yourself to see if this creeps up.
- Write a narrative of how this time of care has been beneficial to you and to the person you are caring for: "We learned a lot from each other. I found I had skills I didn't know. I learned to multitask and delegate."
- Do we imagine the journey of care is only a success if health improves—or is the quality of the journey a better measure of success?

Notes to Myself

CHAPTER 5
Caring for the Caregiver

If caregivers are not healthy, mentally well balanced, and
spiritually sound, then those for whom they care will suffer.
—Leeza Gibbons

The Inheritance: Gael

*I am sitting on this funny little square-backed chair. Today it graces
my bedroom, done up in mint green damask, but once upon a time, it
was the very seat upon which my grandmother rested as her sweetheart
dropped on bended knee to whisper the words she longed to hear, and
soon after that, they married. She told me this on the day she gave me
that chair, lovingly parceling out the few "nice" things she owned as she
prepared for her own death. She added two green jade statues and a
set of wrought iron fireplace tools, and satisfied with her selections, she
told me to enjoy them. Although I knew those moments were special,
I didn't realize until much later what a strong and courageous woman
she was. Her honesty and ability to face life squarely was a lot like that
chair: a wonderful reminder of what was now part of me as well.*

She bore three children, my mother among them, with an elder daughter Virginia, who died at the age of eleven, and her youngest, a son named Raymond. Ray was my godfather, though I called him "Uncle Jim" for most of my childhood. Apparently, the family remained in holdover mode from the days when he wanted to be a cowboy named Jim, but somewhere along the line, he must have decided Ray was a more suitable choice. Along with everyone else, I got used to calling this new-old person "Uncle Ray."

When I was just a girl of six or seven, and he was still Uncle Jim, he shipped off to Korea. God only knows what he saw while overseas in that war. I know what I saw—his frequent place on the living room couch left empty—and up went my homemade tent! Since we all lived in the same six-story apartment building in Brooklyn, I had the good fortune of being able to scamper through the building from the sixth floor to the third floor to visit my grandparents every single day. In the world of the third floor, I was treated like a very special person indeed.

To say that I deserved this is likely stretching it—by a lot. I have numerous memories of tap-dancing my way around the circular living room/dining room, those seductive metal taps just begging to be heard. Sometimes, I jumped on the bed. Beds actually. They had sturdy singles perched side by side with an enticing three-foot gap. What could be better than to jump back and forth, trying out some of my tumbling routines just to sharpen my edge? My grandmother was cringing, yet another year shaved off her life.

When I was twelve, I witnessed my grandfather in the midst of a massive stroke. I could not for the life of me understand why he was sinking off the chair while speaking, his arm dangling limp, and the chaos that ensued. Grandma rallied in circumstances only a caregiver can understand and set to putting him first in all things for the many years following as they lived with his paralysis. It saddened me to lose so much of her. She disappeared in ways that imprinted me with how difficult it can be to hold on to yourself in times like that.

I was visiting her in the hospital just days before she turned the corner in her own life. Dying from cancer and not wanting to "cause any trouble," she asked me to help her get up from her hospital bed to cross the room. As I did, she took two steps and collapsed on the floor. I am still awed by how much presence she had—how concerned about my own welfare in seeing her like that!

Nurses rushed in to take over, and she looked up, encouraged me to go home to my infant son, and simply told me to take good care. I already had the chair, the jade statues that I had once given to her, and the fireplace tongs. She was satisfied with that, I could tell, and she seemed prepared to go on her own journey, neither frightened nor reluctant.

I'd like to think the tongs were a prescient reminder for me to harness my own fire, the jade a special stone that promotes wisdom, balance, and peace. But the chair? That chair! The real gift is found in the story of the love she found—with a follow-up lesson in tenacity that bears witness in us all.

Multiplicity

Sometimes you don't really realize what you are doing until you see it written somewhere else in a formal sense. Gael came upon a list of caregiver skills suggested for a professional resume quite by accident, and this job description included all the skills made explicit for a paid caregiver. The job description for a paid caregiver is listed below. Keep in mind that most family caregivers do these tasks for no pay and without any help. It is no wonder that we caregivers are known for overdoing it physically, mentally, or emotionally. This job description provides a frame of reference for the scope of it if you were to get paid:

- hands-on experience in implementing and supporting patient care plans through exceptional execution abilities
- track record of observing patients to efficiently determine changes in severity of symptoms and report them with urgency

- proficient in assisting patients with everyday needs such as transferring, toileting, grooming, and changing
- adept at planning and preparing meals according to each patient's individual health and nutritional needs
- highly experienced in engaging patients in both physical and mental exercises as part of their health-care plan
- unmatched ability to provide companionship and counseling services to distressed patients
- skilled in facilitating and participating in group activities to assist patients in remaining social and on their feet
- competent at promptly responding to resident care needs and emergencies by employing exceptional first aid (and preferably CPR skills too)
- demonstrated expertise in administering medication according to patients' health-care plans and specific orders of their doctors
- qualified to communicate changes in patients' physical and emotional status to health-care professionals
- deep insight into creating and maintaining a clean, hygienic, and safe environment for patients to thrive in
- highly skilled in handling light housekeeping activities such as dusting, cooking, and laundry
- able to efficiently provide assistance in handling errands such as paying bills, doing grocery shopping, and completing other errands where needed

Take a moment to examine some the job description's key words to be clear in knowing what's actually part of the tasks that both paid caregivers and unpaid family members who are caring for family engage in:

> implement, support, facilitate, communicate, provide, assist, engage, handle, observe, participate, manage, respond, administer, demonstrate, execute, create, maintain, skilled, proficient, adept, experienced,

unmatched ability, competent, efficient, and deeply insightful

What's amazing is how many caregivers who care for family members embody all these qualities along with separate jobs, caring for the person in need and many other family members. Multitasking and handling a multiplicity of roles is how it's done, along with coping skillfully with stress and communicating assertively to ensure needs are met.

Skills for You

Having seen these skills laid out to understand what a caregiver does helps bring to light how many of these skills were not in our repertoire until we started caring in this role. Sharon was a caregiver to parents who both had Parkinson's disease. One key task that was never obvious to learn to do was transferring a person. With Sharon having to lift her parents, she exaggerated back problems, endured a meniscus tear in the knee (requiring surgery), and got a minor shoulder injury from transferring her parents in the wrong manner. Sharon took this more seriously over time by learning how to use a "transfer belt" and learning how proper techniques reduced the risks of falls for her family members and helped her avoid injury. There were also pieces of medical equipment that could help like a lift chair and a pole. Caregivers need to be able to ask for help or use proper transferring skills to make injury avoidance a way of life.

You will need to learn many new things in your role as caregiver. Be open to learning them. Use appointment times well with preparations made in advance to ask the physician. Don't wait until the appointment begins to think of questions (be ready beforehand). Remember that the doctor only sees a moment in time with the patient. Make sure you let them know your concerns in terms of daily care/health matters.

Be ready to enlist the help of the nurse. Many caregiving questions relate more to nursing than to medicine. In particular, the nurse can answer questions about various tests and examinations, preparing for surgical procedures, providing personal care, and managing medications at home. The nurse can make more time to speak with you than a physician might be able to offer. If a matter requires speaking to a specialist to get some type of information, reach out via email or phone. Most occupational therapists and physical therapists can be reached easily and can offer help with a visit or a follow-up inquiry. The answers can't be given if the question isn't posed.

Self-Care Is a Skill—and You Need to Keep It Up

Many times, attitudes and beliefs form personal barriers that stand in the way of caring for yourself. Even though taking care of yourself may be a lifelong habit, with taking care of others, it may fall by the wayside. Think of self-care as a skill you need to maintain. As a family caregiver, you must ask yourself, "What good will I be to the person I care for if I become ill? If I die?" You'll be reminded of the need to manage some elements of your life with a renewed seriousness.

Here is a list of things to maintain for yourself as you also care for others:

- sleep
- water
- fresh air
- walking
- stretching
- food
- comfort
- friendship
- baths or showers
- eating slowing with both a chair and a napkin (not in the car!)
- prayer or spiritual comfort in whatever form you see it

- nature
- listening to music or reading

Some things above are vital, and others may seem optional, but note whether they recharge your energy level. If you want it, schedule it. Also find gratitude for what is. Remember to find joy in those things you enjoyed before caregiving and find yourself "re-finding" them. Make sure to look for synchronicities wherever they may be.

Don't Confuse Self-Care with Selfishness

Do you think you are being selfish if you put your needs first? Is it frightening to think of meeting your own needs? These are typical barriers to get over, especially if you are accustomed to putting others first. Ask yourself what might be getting in your way and keeping you from taking care of yourself and why you see this negatively. Think of blocking out time in your calendar for nothing to be done but allowing yourself the space with no commitments even if you don't do anything. Free time outside of caregiving is time well spent to recharge.

A note about food: We have seen so many family caregivers who "hate to throw the food away" when the person they are caring for doesn't eat it. Don't eat what is left on that hospital tray or the Meals-on-Wheels offering. Don't do it! That food does not contain the same kind of nutrition and caloric density *you* need! Your food rituals need to be sacred stops during the day. Use them to nourish, express gratitude, and enjoy.

Acknowledging Hard Emotions That Come with Family Care

Caregiving often involves a range of emotions—some of which might be ignored during this trying time. Some emotions are intense and can be easy to push away, but the stress shows up in different places when we avoid hard emotions. Signs of stress in your life indicate that you may need to make a change in your caregiving situation by asking for help or hiring help.

Keep in mind that the emotions you are experiencing remind you that you are grieving a loss of what could have been different, yet this is the journey now. Acknowledge that you are experiencing increased stress, and this thought should be a reminder that you need to be assertive and ask for what you need—now more than ever.

There will be hard emotions in this time of care, but there will also be poignant moments of joy. Sharon recalls meeting the ambulance in the emergency room; Gael recalls watching her son finally breathe without the ventilator. Difficult situations help us see the things that count so much more clearly. Those who haven't been caregivers don't have this perspective.

Asking for Help—or Hiring Help

Does the idea of asking for help seem foreign to you? Do you find yourself turning down offers for help with responses like this one? "Thank you, but I'm fine." Many caregivers don't know they need help or are afraid to ask for help.

Sharon once attended a conference for caregivers of those with Parkinson's Disease, and one woman stated, "I can't believe I raised my children in such a way that they don't offer to help me." She was reluctant to ask her kids and expect them to *know* to help without asking. Perhaps you simply want to avoid the "burden" you would bring others if you ask—or maybe you are afraid they will say no. To the person who seems hesitant when you ask, simply say, "Why don't you think about it?" Try not to take it personally when a request is turned down. Remember that the person is turning down the task. They are not turning down you. Try not to let a refusal prevent you from asking for help again. The person who refused today might be happy to help at another time. Either way, admit that you can't handle everything yourself, and something will happen to you if you don't get help.

To begin thinking about finding help for yourself, start by being prepared with a mental list of ways that others could help you. For example, someone might make a visit just to be a companion. You might ask a neighbor to pick up a few things at the grocery store where they are headed anyway. Perhaps a close relative could fill out some insurance papers or look into health matters or options.

When you break down the jobs into simple tasks, it is easier for people to help. If you think people who care for you and your family member won't want to help, you are wrong! They may just need a way to make it manageable. It is up to you to tell them how. Prepare a list of things that need doing that might include errands, yard work, or a visit with your loved one. Let the helper choose what they would like to do.

Other help can come from community resources, nonprofits, support groups, and even paid professionals, although be aware of costs and logistical matters to make these resources useful. Be cognizant that some professional home-care companies with paid caregivers have minimums and may require committing to set shift schedules (four hours, overnight, back-to-back shifts of different people, etc.). If you hire care to come into your home, look for companies that are bonded and insured. Also read Yelp reviews or check Care.com for information on what typical care companies or services offer. Finally, don't be naïve or vulnerable to security risks. Lock up or hide valuables if hired help comes into your home. Although this risk may be minimized with a bonded company, it still is a risk that you might need to plan to avoid.

When you do find friends, family, or paid support who can help you, set up things in places for others to easily find items that support them in their tasks. One friend in Sharon's family filed paperwork related to insurance, noted health-care information for tax purposes, and put receipts in order for tracking. A paid caregiver for Sharon's parents made meals, and the meals were planned in advance with ingredients and recipes left in obvious places. People from church planned dates to pick up Sharon's parents so they could attend Catholic Mass and then brought them home. It may be that a combination of paid and unpaid help is what's in order to meet the needs and not overwhelm you.

The overall lesson is that you don't want to wait until you are burned out and exhausted or your health fails. Reaching out for help when you need it is a sign of personal strength.

Organize Yourself

Charles Buxton is credited with saying, "You will never find time for anything. If you want time, you must make it." This is valuable advice for caregivers who may need to carve out time for organization. Remember that this is an entire topic unto itself, but in the interest of keeping it simple, we felt it was important to address this critically important subject with several recommendations.

First, have a place for everything, and put everything in its place. It's a tried-and-true formula, but you will be served by getting ahead of the curve now. It doesn't matter where that place is! Commandeer a shelf, kitchen cabinet, drawer, or closet. Just do it. This one commitment can save you scores of upsets and wasted time. You can thank us later, but please do it now! Create a dedicated place to put all your newly acquired items, including paper, meds, tools of the caregiving trade, and whatever. Put said objects in their special place each and every time. Let go of having to worry where you can find things when crunch time arrives.

Second, find an organizing method you like. Gael recommends reading *Getting Things Done* (2001) by David Allen or checking out his website (gettingthingsdone.com), which offers simple principles for consolidating things, processing them to handle effectively, and finally regularly revisiting your to-do lists. There are scores of organizing principles out there, and if you are feeling like life is chaotic and overwhelming, having a framework for managing thoughts, tasks, and daily to-do's will help. If you have begun caregiving, you are already at the stage of needing this help.

Third is an idea many people swear by. Try to touch everything that is nonessential only once. If you have junk mail in your hand, recycle it immediately. Don't bring it in the house and put it on the counter. You already know it is junk. Deal with it so you don't have to touch it twice. If you have a container of milk in your hand, put it back in the fridge right now. Try to realize in a typical day how many nonessential steps eat up time and energy.

Fourth is to begin a two-minute rule, especially if tasks start to pile up. If you can do it in two minutes, do it now and don't put it off. Sharon has an hourglass she uses to begin tasks that are quick and need a speed challenge. Some tasks might be big—like cleaning—but when you break them into pieces, they can be manageable. If you usually use a mop with a bucket and now you use a Wet Swiffer-type mop product that you can throw away after mopping, you may save yourself time and get a task done quickly and more easily than in the past.

It may be that you don't need more time, but you may need to use the time you have more efficiently. See how you can use these ideas in your own way.

Chapter 5 Reflections: Caring for the Caregiver

- Make a preemptive list of deal-breakers you know you must keep in place in order to maintain your own sanity and well-being going forward.
- Practice letting go: Reconciliation happens when you know you did your best. For you to be the caregiver you wish to be, what would do have to do? Forgive someone? Forgive yourself?
- What is the most important area for you to organize right now? What baby step will you take today? Make a list of what else needs to follow.
- If you could time travel and speak to your old self, what would you tell them about how to thrive with what's ahead?

Notes to Myself

CHAPTER 6
Caregivers at the Center of It All

Caregivers are often the casualties, the hidden
victims. No one sees the sacrifices they make.
—Judith L. London

Camping: Gael

*Popsie (my grandfather) and Debbie (my neighbor) were two pillars of
friendship throughout my childhood. Debbie, the angelic girl next door,
was just shy of three months older than me, which means a great deal
when you are only five. She and I went to the same playground, played
the same games, went to the same school, sang in the same church,
and endured the same mean ballet teacher. With so much the same,
those three months were a chasm of difference that gave her status in
meaningful ways I could never catch up to.*

*Then there was my grandfather on my mother's side, as I
remember him before the stroke. A retired cop from the streets of*

Brooklyn, he was a handsome man who looked as dashing in his dress blue uniform photo as in those from his lifeguard days. He shared lots of things with me, but mostly things he loved, like the racing form and gin and tonics. To my mother's chagrin, we were coconspirators in most things. He, a strict disciplinarian in her own life, was the devil in mine. I thought it was wonderful! Field trips to Freedomland and Coney Island were a regular part of those good times. So were summers at Spring Lake, where my grandmother would make us cream of chicken soup.

One summer, he surprised me with a gift from the Rent-a-Tent Organization, the outfit Uncle Ray represented for a time. After filling the trunk with all the necessary gear, off we went on my very first camping trip. I was beside myself with joy! That experience sparked my interests in permanent ways and profoundly shaped me into the outdoor adventure guide I eventually became. Would that he might have enjoyed it as much himself. He was clearly done after only one night. Nonetheless, the die was cast for my future excursions as we packed up and headed home to the couch, and I continued to dream on about tents.

How does one manage to take a dynamic, impish person and fit them into a paralyzed body? Throughout my young adult years, he rallied from death's door—but never again back onto his feet. The amount of time and care he required was astonishing as the management of his condition became the defining structure in our family orientation. I bore witness to his incomparable ability to express himself through new interests, which suddenly included mosaic art.

My mother and grandmother began a life of new considerations for his care. These had far less to do with art and everything to do with hard work and service. I saw the intricacy of connection and the inevitable need for balance when what lies latent in our caring nature stretches to unfathomable lengths when we are called in whatever ways we are to be responsible for another through it all.

Caregivers Are Not Always Perceived Well

Family members are a presence, and their involvement to support and aid or even steer the direction of a family member's care is often critical when the person getting the treatment can't do it themselves. However, attending a doctor's visit or a therapy appointment with a family member does not always mean a positive experience. The caregiver is not always a welcomed or included person, and voicing an opinion about care can be seen as second-guessing it.

When Gael was speaking with other Caregiving Consultants on how family caregivers can be seen as the valuable assets they are, Leslie Cottrell Simonds, author of *I Want You to Know* (2016) stated:

> My first thought comes directly from feelings shared with me by practitioners that are open to a happy healthy relationship with patient and caregiver. However, many are leery of being inclusive of family caregivers within a professional circle of providers ... many professionals see caregivers as uninformed, opinionated, and frankly a thorn.

Imagine being seen as a thorn! Cottrell Simonds notes that family caregivers can be seen as quick to demand, and this can feel like overreaching by medical staff who may be burned out or overwhelmed with their patient load or the complexity of their work. To combat the negative perceptions some medical staff may have, she recommends that caregivers should be ready to be engaged and supportive with staff and be humble and willing to learn. Being educated and informed is vital, but it is important to be careful with the ways in which questions are posed. Having the facts first and keeping an open dialogue are critical in keeping options open:

> Part of the paradigm shift of the new age we live in requires that we all learn to remove our [so-called rose-colored] glasses and put on the glasses of the

person or professional sitting across from us. As we teach caregivers to step into their power as educated, loving, and involved components of the "whole" of care of their loved ones, the perception of medical professionals will definitely become more inclusive.

Central to these themes is that by using a sense of realistic understanding and pushing aside the desire to avoid things that are hard, a caregiver can begin to judge a situation with more clarity and begin to assert a strong advocacy voice on behalf of the loved one they care for.

Later in this book, we will discuss further what it means to consider yourself a "hub" for the set of experts involved with the care at hand. Having a role in this circle of care means knowing your voice and having it be heard. There may be areas in which a caregiver and the family member being cared for want a second opinion, seek a complementary approach, or have concerns about side effects of a new treatment or drug and ask for explanations. Being a part of the team requires being able to assert the family member's wishes in ways that may not come easily to all family caregivers, but these skills can be learned.

Skills for Being Assertive and Speaking Up

Assertiveness, which means for our purposes to be stating an opinion openly yet respectfully, may be confused by some as aggressiveness, which can be about attacking the opinion of others. Keep in mind that it is the job of an advocate to speak about someone in your family's needs, ideas, or wishes in a careful way. It is most easily done with well-intentioned ideas and a calm, well-reasoned approach and delivery. Although aggressiveness is a more forceful approach to getting one's way, it can be received poorly by others. But if a caregiver is easygoing and not wishing to make waves with the medical staff overseeing a caregiver, they can be assertive and heard without prompting a staff or doctor to feel defensive. Assertiveness means not being afraid or timid when expressing an idea or need that you wish to be heard. It involves capturing the attention of another and ensuring your idea is heard. It may also mean aiming to negotiate and being willing to give and take.

In some instances, a caregiver needs to assert themselves to speak up for the rights or comfort or conditions of care of those they care for. In the case of traumatic injury, Gael found that inquiry beyond blind acceptance of parts of the radical—and experimental—care plan for her son's prognosis was significant, and she had more choices than she was offered. Sharon also had care of her elders with geriatric care and her son for developmental delays. In both cases, a set of steps could have been accepted without questioning it further, but by asserting questions and asking for more options, a dialogue opened up for expanding options on treatments, care, and involvement of others, such as therapists, which were not initially posed.

Using Outside Advocates

Sometimes using an outside advocate is worth it. Sharon had experiences that involved advocacy by a free county Ombudsman when she had to raise a concern to a higher level about a senior facility's seating arrangements. Sharon's parents were both struggling with Parkinson's, which affected voice and ambulation, and they were less vocal about their needs. For several weeks, they complained to the staff that they didn't want to sit with only those at their table who spoke another language, which meant no socialization went on.

In spite of her parents' complaints, management and staff did little. In fact, the director said a new seating arrangement could not take place because it would be disruptive to the longtime residents. To escalate the matter, Sharon was able to use an ombudsman whose role was to provide unbiased oversight and intervention in senior facilities. Their conversation about seating arrangements led to a seating change and other beneficial changes in Sharon's eldercare plan. None of these plans for anticipating setbacks would have been possible without the conversations with an outsider who guided Sharon's family though the harder and easier aspects of care. Assertiveness and outside pressure mattered.

Geriatric-Care managers and Certified Caregiving Consultants are among caregiver supporters whose role involves payment, which can allow the family caregiver to step back from making all the hard decisions in a vacuum. Even Medicare advocates can be useful, and their assistance is free of charge in providing guidance on understanding billing, claims, and even advice on making appeals when a person feels billing was in error. These external sources can be available to advise or communicate directly in advocacy situations, depending on their role. In whatever way the advocate is used, be mindful of contracts, hourly costs, and privacy factors. An outside expert or advocate might need to be used in a formal role to have them involved in direct decision-making aspects or providing guidance and advice without acting as an agent.

As you find your voice within the caregiving role, it takes clarity and guts to be real and get to the point. Gael said, "I said no to authorizing experimental drugs for my son, and insisted on an alternative option for my mother when simple solutions were right in front of everyone. There were times when, in the face of a system with so much reliance on modern chemistry, the only one to point to an alternative was one of us outside the system."

Trusting your intuition and using your voice to question available options will serve you well—even when the options are unclear and require probing and pushing to get answers.

Supporting Caregivers as America Ages

As caregivers, we support, we bear witness, and we accept. It's a presence needed during any time when a health matter arises, but predominantly, it is felt by families whose members are aging into their elder years. What used to be "old age" a generation ago holds a different age marker now with elders now living into the hundreds of birthdays. With more years lived, more health issues arise, for which caregivers are needed to attend to with the familyo member.

In 2016, the National Academies of Sciences, Engineering, and Medicine produced the "Families Caring for an Aging America" report. In accordance with their mission "to provide independent, objective advice on issues that affect people's lives worldwide," the report stated that by 2020 as America ages, 90 percent of all working adults will be affected in some way at work by organizational responsibilities and time spent in unpaid caregiving. The financial effects alone are likely to become as cataclysmic to organizations as they currently are to individuals. The simplest and most valuable solution is the one that helps the caregiver function effectively and efficiently, which means helping caregivers improve communication skills, acting in the best interests of their families, and being a force for those they care for and themselves.

Caregiving is not easy, especially when the role is unpaid, which is often the case in a family-care circumstance. In addition to the lack of remuneration, this family care also creates a time burden, which those who are not caregivers do not recognize. The report also stated, "In order to fulfill the roles that they play, family caregivers must interact with a wide range of providers and navigate within a variety of systems. They interact with physicians, physician assistants, nurses, nurse practitioners, social workers, psychologists, pharmacists, as well as physical and occupational therapists, direct-care workers, and other noncertified but medically related staff." It is known that caregivers already serve as key information sources about health and past histories, but they also have to also keep careful notes of medications taken (and possible negative reactions to drugs), as well as previous treatments and ones to be taken advantage of, or avoided. In addition to being carefully attuned to tracking details like medications to convey, caregivers also can influence care decisions as advocates and facilitators. Caregivers are crucial in offering ways to benefit the wishes of the person cared and this is especially true for those cases where the family member cannot do it for themselves.

Due to the nature of the close relationship family caregivers have to the circumstance, their actions help alleviate problems that may arise when only strangers are involved in a person's care. Caregivers are especially powerful to help manage and support care within formal networks and within as well as across the family spectrum. This participation by family members tends to be a tremendously undervalued role, and it often is minimized even by those who work in professional health-care settings. Knowing how caregivers are able to make their presence known can work to overcome this marginalization if it should arise in a health-care setting.

If family members know they may face barriers in the course of caring for their family member by those who undervalue their presence, they can take actions to be a forthright, proactive voice in a care plan rather than a dismissed presence with a defaulted passivity. This means a caregiver must actively communicate and engage with others not as an overbearing presence but as an effective partnering influence working with the formal medical and nonmedical members of the care team.

Overcoming Caregiver Invisibility

To be seen and not valued is not only limited to doctor's visits. Being a caregiver and not knowing others in the same situation can be disheartening and lonely. Even being one who attends support groups feeling like part of the audience and not a participant can make some caregivers feel invisible. There are even those who don't call themselves caregivers. To be separated from others whose plight is similar—even if the person cared for has a very difference diagnosis—can cause an increased sense of burden.

Caregiver Support Groups

Caregiver support groups are not a new phenomenon, and many types of medical conditions have group meetings and special times and designated circumstances where caregivers can meet together. Many caregivers attend only disease-specific support meetings, but their needs are not always met at these meetings, which focus upon the person they care for. A better option for some caregivers is to attend a nonspecific caregiver-support group. These are perhaps less known since people might think there is less value in gathering with people whose situation doesn't mirror the conditions present in their home.

Caregivers share so many similar experiences—even if the people they care for have different symptoms, problems, and durations of care. Finding these support groups aimed at helping caregivers in general are important to research through online forums or physical meetings in nearby locations that are convenient to attend. Many churches have meetings like this, and even online meet-up groups have informal gatherings that help caregivers connect.

What caregivers know and can share together, even if their family members have difference diagnoses, is that care is hard. Being heard by others who also struggle with priorities and multitasking is helpful. When caregivers admit they need to change to those in similar shoes, the sense of obligation is different. For example, being accountable to others who know you are being stretched too thin can allow a caregiver a sense of permission to hire care when non-caregiver friends may think it is an unnecessary need, when in fact, the delegation of the care may mean a family caregiver gets needed rest and time to regroup.

Sharon was attending a caregiver-support group meeting of persons whose family members had different health problems. It was the first time she attended a meeting that did not focus only on Parkinson's, and the symptoms fellow caregivers described matched hers: exhaustion, frustration, resentment, and a sense of falling apart. To have been at this meeting was to have found a tribe of like-minded people who understood what it was like to fight, advocate, and even accept hard situations with grace. It was like a breath of fresh air that offered benefits for her own well-being and health.

At other meetings, Sharon gained tips on medication-management logs, advanced care directive ideas, and interfacing with family members living far away to promote communication and strengthen family participation at difficult times. Sharon even took what she learned in this situation of a face-to-face caregiver-support group setting and started her own online meet-up group with the same tenets of a physical meeting but with the presence of an online support community.

When caregivers acknowledge and cultivate their own role with other caregivers, it lightens the load mentally. It also helps them manage the roles by learning from others' experiences, including how other caregivers engage in their circumstances. Speaking with other caregivers in privacy and in the comfort of a discrete setting can help caregivers commiserate and find renewed empathy and resilience. The sharing among caregivers doesn't need to be specific to a disease type. Rather, it is the commonality of caregivers in similar support roles that proves invaluable.

Getting Your Own Needs Met

When a caregiver does not allow for time to attend to their own mental or physical needs, health problems can emerge. Grant Charles, PhD, an associate professor in the School of Social Work at the University of British Columbia, studies how caregivers can suffer if they put themselves last.

In a feature article in the American Psychological Association's *Monitor on Psychology* (2015), Stacy Liu wrote about "Invisible Caregivers." She cited Dr. Charles: "There's nothing wrong with having to provide caregiving ... but there's something wrong if you're not getting support yourself or your needs are not getting met in the process." Dr. Charles's experience and insight is valuable in underscoring how caregivers must acknowledge and embrace their own care in the midst of caring for others. Knowing that caregivers' needs won't be met unless they are assertive in stating their needs and meeting them raises the bar. Sometimes only like-minded caregivers can understand this point of view and are the best people to be accountable to when attempting to create beneficial life goals after a period of self-neglect.

Certified Caregiving Consultant Vivian Geary reminds us about a hidden toll on caregivers who can simply forget to take care of themselves when morning routines get busy. "Because caregivers hit the ground running, they often sacrifice self-care rituals as the high incidence of periodontal issues attest." We can all relate to one thing leading to another as the day progresses, until just remembering to brush your own teeth can take a back seat to the responsibilities at hand.

Keep in mind that not everyone understands caregiver stress. This may even include those who are in close proximity to caregivers and those cared for such as physicians, nurses, and other staff who might not always get the need for programs to support this population and ease some of their stress. Another aspect affecting caregivers is the need to be vocal about personal and private issues. For a caregiver who is struggling with balance and self-care, to voice these concerns aloud might mean becoming vulnerable to public scrutiny and feelings of discomfort that could have been avoided by not speaking up. Feeling reluctant to talk about the emotional, financial, and physical challenges of care means caregivers would remain isolated, even among the very people they often look to for help: the medical community.

When the Medical Community Is Dismissive toward Caregivers

In addition to isolation, some caregivers feel their role is often dismissed by medical professionals. Although the "Families Caring for an Aging America" report found the presence of caregivers to be useful and essential, they also found that some medical staff don't accept the caregiver's presence but rather show disregard and lack the time necessary to work with families. The report states, "Family caregivers are often marginalized ... some providers exclude them from older adults' treatment decisions and care planning while also assuming they are able, have the knowledge, and are willing to perform essential tasks."

Physicians and staff who rely upon family caregivers to follow through with medical directions once the patient leaves don't benefit unless the caregiver is involved. The lack of training and attention to this critical audience leaves room for errors in following medical instructions at home since "caregivers describe learning by trial and error and report a fear of making a mistake." Family caregivers need to be at the bedside for patient instructions and to have their concerns and questions heard and addressed. To avoid this interaction or marginalize the caregiver helps no one and may worsen the patient's condition and experience, thereby making the life of the caregiver harder.

When working with medical staff, it's important to learn as much as possible about the medical situation and procedures. Staff and medical personnel appreciate when you can demonstrate some understanding or are willing to learn all you can. Be prepared to keep good notes about advice, discharge information, required and optional therapies, and post-procedure actions. For specialists and referrals, always get copies. If needed, ask for dates, numbers, and prescriptions to be repeated or so you can double-check. Also know that you can share observations in your role as a caregiver. Knowing what you are expected to do so can be a shifting experience with more involvement at certain times than others. You are entitled to ask or observe or question—don't let the circumstances be daunting.

Chapter 6 Reflections: Caregivers at the Center of it All

- Are you making time for self-care and if not, why not?
- Do you have things that are pent up that you pour out on others unknowingly? What good would come for you without self-care that has to change now?
- What can you add to your life to help you grow and feel rejuvenated?
- List ten items that are non-negotiable. How can you implement these in your life—even in a phased approach?

Notes to Myself

CHAPTER 7
Acknowledging Our Roles

Doctors diagnose, nurses heal, caregivers make sense of it all.
—Brett H. Lewis

I Did It My Way: Gael

It has been seven years since my father passed. Today, I went online to view his legacy page, a placeholder my family created for people to upload notes, light symbolic condolence candles, and remember. I noticed the war buddy comments that were there—it was mostly war buddy comments actually, those of them who were left. Those connections had always remained current for him, bringing emotion and relationship to the 1940s more vividly than a lot that happened after. They were a band of brothers, and World War II bound them for life.

Their comments revive a clear memory of one of the days we spent at the rehab facility attending Physical Therapy sessions when there was still a modicum of hope he might get back on his feet after yet another hospitalization. I am the visitor watching while the physical therapist offers exercise instruction to the small group of patients in the room. Each in turn receives her attention. "Ten repetitions here and another ten over there!" she calls out. She is enthusiastic as legs are lifted and ankles curl. This is work that will enliven his legs, the therapist explains as needed encouragement, and "perhaps improve his balance" as he, like a kid who enjoys skipping school, practices passive resistance mostly.

Tick tick tick. Just not gonna do it.

The session ends, and back in the wheelchair we trundle to return again another day. Frustrated, I ask him what gain could possibly be had of such non-doing.

"Set the prisoners free" is his studied response, telling me again of the infamous Bataan Death March of the war and the perils of too much exercise. As a movement therapist myself, I realize how thoroughly captive he feels inside this body that no longer works.

I can only hope to recognize what actually served him in the end. After all manner and form of support given, our family came to know that just letting him be was what he needed—and what he demanded. I was with my mom some weeks later when he defiantly refused to go to dialysis, the final of three strikes signifying you're out. There is not much time left after that. Simply, the die was cast. Hugging my sobbing mother while she surrendered yet again to the transport van leaving without him, this too defined a caregiver's role for us both.

Taking Your Place as a Modern Caregiver

As a caregiver, you are not alone. If you look around, you will likely find an unpaid family caregiver sitting patiently in the waiting room of every medical office in the nation. Caregivers are the unseen backbone of our care system today. Unpaid family, the friendly neighbor, and the good buddy are all on the front line when an individual needs support and can't provide it themselves. There is the help from a son or daughter driving a mother to a doctor appointment, the delivered meal when a neighbor is hospitalized, or the ferrying back and forth to routine doctor appointments that can be done by friends—all of these unpaid and welcomed. But for a primary caregiver—the main person doing and handling or scheduling or assigning tasks, caregiving is so much more than transport, delivery, or appointment making.

Gael's father used to say, "Growing old ain't for sissies." Neither is becoming a family caregiver, especially when you're doing it over a long period of time. Caregivers are part of the team of people caring for a person with a health condition. This multifaceted group of people supporting the patient is headed by the medical staff who direct the care and its formal follow-up care, yet the follow-on care that happens at home has oversight by a caregiver. Even when family members are residing in a facility for short- or long-term durations, caregivers assist, explain, and experience matters related to care closely and personally. The caregiver is the everyday slogger, often the most in tune with the person cared for, and the person who sees them as more than a patient.

Things are rarely simple with family caregiving. Complications and medical complexity happen with learning curves. The family member may be on multiple medications, have many specialists, or require in-home care for hard tasks many families have no familiarity with—changing catheter bags, providing injections, changing bandages. There's no room for sissies there.

Just because a family caregiver is assigned to take on a task to medically help a family member does not mean they are expert at it. There is learning and growing into the complex caregiving role. Part of this role means caregivers see when something isn't right or when a medication's side effect or adverse reaction is taking place. Because of this vital importance of the caregiver, their presence must be heard as they take their place among the care team and use their voice to powerfully interact to help their family member.

Finding and Using Your Voice to Advocate for Your Family Member

Speaking with conviction about your own needs and the needs of others you care for may not come naturally. For some, it requires learning and persistence.

At a residential care facility where Sharon is an Ombudsman advocate for elders, a woman said, "I've never been one to stand up for myself, yet I heard myself do it—and I didn't waver. I don't know where that came from! I never did it before, but I am so glad I did." That woman learned to self-advocate in her eighties, and she was awed that speaking up for herself could work. There is no need to have born-with-it skills to advocate for yourself or family—just be open to many of these notions explained below and practice using them with persistence.

Assertiveness skills are advocacy skills. To assert something is not just loudly talking; it's talking so others will listen. You will use voice inflection and tone better if you use it to your advantage and not to overtake. Body language and other keys to effective communication are outlined in this chapter as well as ways to see how you can emerge differently as an assertive caregiver at the reins. Two more vital aspect of assertive communication are authenticity and integrity.

As many caregivers know, there are times when standing up for a family member means going against the grain. During an online caregiving summit, Sharon gave a caregiver talk on the topic of knowing the right time to move a family member to a residential care living environment away from home. It was a tough topic and one which is often associated with elders'worsening health when living at home is no longer safe or the care at home cannot be upheld continuously. After the summit was completed, Sharon was contacted by email from a family member who had to place a young child in an intermediate care facility (ICF) and also spoke to a mother whose young adult son with special needs was in the process of finding a group home. No matter the age of the person moving from home to a facility, the very idea of moving a family member can cause the caregiver to be deeply conflicted. The decision of where to have a family member live when home is no longer an option is a topic fraught with indecision and guilt. For those caregivers making these decisions, they must come from a place of understanding that home is no longer an option, and persevere. Looking for the right fit in a care facility is only the start of the work and it's something to be done together with the loved one contributing. Keep in mind that once a facility or group home is found, advocacy for a family member living in a new location will be required. The role will require staying in touch with care circumstances and ensuring that staff are upholding dignity and capability at all times.

Communication Fundamentals in Advocacy

Some guiding words from Gael can help you find your voice as an advocate: "Say what you mean, mean what you say, and don't be mean."

There will never be a better ground rule or guiding philosophy than this for a caregiver. Advocating for another person is best when it comes from a place of personal understanding, knowing your own influence, and effectively stating with conviction what needs to happen for the person you are caring for. Integrity and doing things you feel are right and sound are at the core of being assertive. Acting in a way that feels wrong to you or oversteps your own ethical bounds may mean you need to reexamine this role or reestablish the bounds with which you will take on the caregiver duties. You should not be compromised.

Sharon was once in attendance at a caregiver seminar for those with Alzheimer's Disease. In this seminar, the leader said, that in the caring for a person with memory issues, there will often be "the need to lie" and that this lying is okay. In Sharon's words: "It just struck a nerve with me to have it be implied that it might be okay to lie casually." Sharon explained that in her present three-generational household, there is a family member with dementia. With so many people's needs to meet in her household she states, "The idea of lying in our home to placate one family member was something I'd hope to avoid as we have younger kids watching and learning from this experience." There are times when white lies help and move things along when a memory-impaired person is "stuck" but it is another to advocate lying where it may become an ingrained part of life.

Sharon lives in a household that crosses generations and ages from elementary aged children up to octogenarians. To have one set of rules for caregiving an elder and one for kids is confusing. Thus "say what you mean and mean what you say" might take more work to get the message across without lying, but it involves more intentionality and avoidance of lying. Sharon has found that there can be success with honesty even if it means working through unique avenues. No one will have to "remember which lie" to keep up with if honesty is the rule of the house.

Use Body Language Effectively

A key to effective and assertive communication as a caregiver is to be aware of body language and the dynamics from you and other people. We all know how necessary eye contact is (and how unhelpful eye rolling is!), but body language is so much more than that. Physical responsiveness to another through your presence will help you to talk so that others will listen. For example, you can lean in toward the person you are speaking to. You can gesticulate to show emphasis to help you communicate clearly.

Be aware that you might be bringing passion to the conversation and moderate your emotions. Be cognizant that any emotional response that may come across too strongly might minimize your interaction and lessen the chances of you getting what you need. Watch your body language and remain approachable even if you are not fully cooperative with the ideas discussed. Always look others in the eye, speak with clarity, and be ready to listen with interest to their replies or concerns. Facial expressions trigger corresponding feelings, and your unconscious body language and countenance can alter the emotional state of the person you are speaking with. Try to remain in touch with how you are conveying information to others.

When someone unconsciously imitates your body language, it can be their way of nonverbally saying that they agree. Being aware of your presence and acting and moving with intention can be important parts of building feelings of mutuality with others. Another way to make the other person feel understood and accepted, which doesn't even involve words, is to gesture to make your point. Since experts say that gestures are integrally linked to speech, gesturing as we talk can actually power up our thinking. If you are not a person who uses gestures, try them. See if the physical act of using your hands to make a point helps you form clearer thoughts and speak more declaratively. Speaking and acting assertively takes time, but even if it feels unnatural, don't give up. Keep on trying.

Make Time to Prepare

Some interactions as a caregiver require preparation. Whether it is a hard conversation with a doctor about transitioning to hospice care or finding out about coverage limits in advance of decisions to seek a treatment, allow for uninterrupted preparation time to examine the course of action so you are able to act upon the necessary decisions with the right information. You may need to set the condition of having a kind of mental equanimity to consider the harder options that worry you or scare a family member. Allow for a place of quiet and a phone with the ringer off so that you can give the necessary focus to the preparations you will need.

Don't expect that there will always be time for a discussion at the end of the appointment or interaction. You may need to schedule an appointment with someone rather than having a casual conversation in the hallway or reschedule when they have a moment. Advance preparation is essential, and if you expect to have to challenge something, be ready to plan for resistance and be able to counter that.

For example, in the situation in which you are planning to challenge a set of opinions or advisements by a professional, be ready to research and know what others have done. If there are family members who can help or advocates in your community (like an ombudsman for a county supervising residential care facilities), use them if you need them to help you.

Plan ahead to be prepared to say what is important to you—even if you have to script it and keep notes with you or practice in advance. The following are areas to consider when you are asserting yourself:

- Check your breath routinely in conversations you are nervous about. Try it now by slowing your breath. Try to lengthen each exhale for a few breaths and watch how your nervous system calms. Use this to ground yourself in the car or before a tough appointment in which you will need to be mentally ready for a discussion.
- Remember that "No!" is a complete sentence. So is "Yes!" If you need to come back to someone later with further specifics or justification, just say it. Managing the expectations of others is a powerful way to postpone and allow yourself time to come back.
- Seek the simple solution—and don't be afraid ask for it.
- If you feel it is important to say no to unnecessary trips for health care, automatic doctor visits, or tests, then question these options or ask for alternatives if they are possible. The medical community is always ready to "do more," but if your family member can't or doesn't want to do more, help that voice be heard.
- Be aware of what insurance will cover and what might be out of pocket. This may mean several phone calls or emails, but make time to learn this vital information rather than being upset at the end of an appointment in which you are taken aback by costs you may have to appeal later on.

Be clear on your talking points—and be ready to advocate with backbone and from a place of authority. It helps one know what is true apart from the fluff that surrounds it. One example to elucidate this kind of advocacy is when Sharon's mother was losing her voice to Parkinson's, and Sharon advocated for a protocol named "Lee Silverman Voice Treatment," which is a set of speech-therapy appointments aimed at having the patient practice tones and articulations that save the voice from being lost entirely. Sharon spoke to the speech therapist and was told it would not be covered by insurance. Research into out-of-pocket payments for the treatment was followed by an appeal process to the insurance company. Oddly enough, the insurance company agreed to the appeal—but only if Sharon would drive her mother to another county that was more than two hours away for the four-day-a-week appointments. Of course, they knew it could not be done from a practicality standpoint. Through steadfast advocacy, Sharon was able to work with a local speech therapist in the insurance plan to do "modified" Lee Silverman-type practicing of speech and sounds. This met the need to continue the voice practice—but not with the very strict protocol the other program could have done with more benefit. The end result was a compromise that was better than giving up or driving too far for a treatment in another service area. It was a big win and an illustrative example of how prolonged advocacy creates results even in unexpected ways.

As Sharon's example with the speech therapy can show, sometimes the decisions you will encounter will be hard to evaluate. For Sharon, to know that her mother would no longer have a voice was a hard reality to accept if a medical speech treatment could not be found. For others, advocating for a family member to stop dialysis, which means certain death, is also not an easy decision to support when you want that family member around. Be aware of the difficulty of these topics. Be present with your feelings when you have to consider options. It can also help bring clarity to sort through your feelings with a journal or a discussion with another person. Also writing or saying fewer words—writing a bulleted list of pros and cons or creating a haiku for the poetically minded—might seem hard to do, but the results are meaningful and quickly get to the point. When you are offering an opinion or point that might be difficult to receive or might be initially refuted, be ready to say what you mean and stand your ground.

If you can, find medical or supportive staff who can stand behind your hard decisions. Sharon found an ally in the speech and language therapy department who was willing to help in a novel way with a new treatment. For another person whose family member meets the criteria for hospice, a physician or social worker might explain the "readiness" for hospice to those in the family who are not prepared to hear of the impending journey toward death for a loved one.

Sharon once had a conference call with family members and the physician in which it was stated that "Sharon was pushing Dad onto hospice" with subjective and fear-filled implications being felt by all hearing those words. The physician was quick and ready to respond to the delicate situation. The physician was acting as an advocate for Sharon's parent to die with dignity and was able to guide the conversation from a place of discussing death to focusing on hospice-readiness factors. By using this medical framework and not emotionally charged phasing seen as "pushing" or "forcing" words, the decisions at hand were easier to discuss without getting caught up in the emotional language of fear.

How Your Way of Being Will Change

Many caregivers find their voices as they move along the caregiving journey. Assertiveness need not be mistaken for pushiness or aggressiveness because it involves an even temper and a clear sense of boundaries. By contrast, aggressiveness overtakes boundaries and seizes on control and anger. Assertiveness provides clarity and insightfulness that begs to be heard and understood. Assertiveness involves gaining an audience by using words that are carefully considered by being willing and caring enough to make points that are well founded and loud enough to be heard and not overrun with the reply.

Some of the personal areas that change for caregivers involve finding strength in unexpected situations when they challenge elements in family structure or medical models they are dealing with and supporting. You will be ready to handle topics and situations that include:

- acceptance (also known as "laying down the sword" of hard decisions)
- managing hurt feelings of others
- misplaced righteousness
- knowing when others or even you have a "needing-to-be-needed" response
- recognizing when "grasping at straws" is at play
- managing hope and addressing fear

Your advocacy may or may not always be successful in every situation. There will be instances of rejection, but you must be the one to take things in stride. Ask for second opinions, check with others who have the same conditions or doctors, and be ready to accept news or options that can't be changed with grace, keeping in mind the dignity of the person you are caring for is of utmost importance. Resilience skills are vital for those receiving care for challenging health news, and are always important for caregivers who are faced with an ongoing sense of needing to be the person managing the next steps.

Avoiding Aggressiveness

People often confuse assertiveness with aggressiveness, but they know it is not passivity. What is the difference? People who state their opinions powerfully yet with respect of others are assertive. However, aggressive people are known to attack or ignore what is said by others, always turning the discussion in favor of their own ideas. By contrast, those who are passive don't state their opinions at all; therefore, their ideas are never brought to the fore. Using assertiveness means acting respectfully and engaging others without rude or dismissive behavior. Knowing the conditions in which you will be communicating is important for assertive communication.

Assertive communication is done best when the person acting is aware of the power structure and influences of the circumstance in which a change may or not come about. Sometimes it is best to know whether an idea you are asserting is a long shot or a reasonable possibility. Either way, if you are new to being assertive, you are making yourself heard and not keeping things in (like a passive person would do). If you feel this is new or hard for you, write down what you want to say. Be ready, practice, and rehearse it if needed. You might find you are overwhelmed, and if so, it is not the time to be assertive. Take time to prepare for and schedule another opportunity to speak up or use outside channels to influence your actions. Don't give up.

One way to ground yourself for challenging circumstances that require assertiveness is to revisit the balance in your life. When you are overwhelmed or depleted, advocating assertively for another person will be a challenge, but it will be easier coming from a place of calm.

Working with Others in the Family on Care

Why do family members who all have the goal of giving good care seem to fall apart when the going gets tough? There are often stories of families coming apart and fracturing over caregiving responsibilities, and the instances of siblings having problems seems to far outweigh the instances where things go smoothly. Sharing information is a form of caring.

You may need to suspend the idea of care issues as separate from family-care issues for a moment and recognize that any effective group needs to work like a well-honed team in order to work together well at all. One of the ways a person can unconsciously control the position within the family is to control information. The closer family member or primary caregiver often knows far more about what is happening on a day-to-day basis with the person being cared for and can choose, consciously or not, to share the information and consideration with their more distant siblings—or not. The sharing offers the possibility that siblings can step forward and help where, when, and however they can without judgment, direction, or command.

Emotions, Problems, and Boundaries

While we may come to terms with the fact that we can't make our loved one well, we still want to be the person to provide care and protection. It's a protective instinct that's hard to overcome. Working to know why we are feeling a certain way or why others with whom we interact are responding defensively can be helpful for gaining insights into a situation and addressing it better. Below are some areas where emotions can create problems:

- Guilt: Sometimes guilt can enter the picture. Though it's often not recognized and most of the time is undeserved, one may feel that we could have done something to prevent what has made our loved one so vulnerable. Perhaps our position as a spouse, adult child, or even a parent requires us to personally provide all of the care needed. Deserved or not, guilt is nearly always a useless (and sometimes destructive) emotion, yet how to better manage guilt is a common problem for caregivers.
- Competition: Some caregivers may, especially in the case of adult children caring for their parents, still be trying to earn our places in their hearts as the one who did the most. Sibling rivalry, even in healthy families, seldom completely

disappears. Some would give an arm and a leg to get any help at all from siblings or at least to keep the siblings from criticizing their own caregiving methods while offering no constructive assistance. For others, there are caregivers who shut out other family members. Most likely, they subconsciously want to be the family hero or martyr. We've heard from enough shut-out family members to know that this touchy subject needs to be addressed. Engagement can be possible even for those who are far away. Delegation and letting go are powerful ways to share difficult caregiving actions, but they must be thought out and executed with a plan made in advance and all buying into it.

- Fear: Many don't trust hired caregivers when we feel we can do better, but this can take a toll on the family and may need to be reconsidered. Whether hired help are providing in-home care or respite care is given at an assisted living facility or a skilled nursing setting, paying for care can reduce the stress and toll on a caregiver. All of us have heard horror stories and may even personally know people who have had terrible experiences with hired care. Being fearful of these rare situations makes us afraid of what may happen if we are not present to monitor our loved one's care at all times.

Good communication helps create proper channels for feelings to be heard. It also allows people to feel closer to one another and feel as though they know what is and is not wanted because everyone is able to communicate that. As a result of communication, improvement can be felt with proper boundaries being set. With these skills, you will have more opportunities for clarity in all your relationships. There is no reason why people should not focus every ounce of their energy on this one important task.

Engaging family members (even extended family) means sharing what is occurring and what is needed among family members is essential. At the start, you may not realize just how much is going to wind up on your plate as a caregiver. You may think you don't need help or consensus in making decisions—until you do.

When you document and share the landscape of what is occurring, and eventually appointments and to-do lists, you allow family members to step forward and do what they can from afar. A sister or brother living at a distance can take on a call to an insurance carrier. Another might step forward to organize online ordering of supplies. Another might need to plan airline trips in a way that only they can do with as much information as they need about placing their visit among ongoing medical tests. Transparency helps you share in a way that is helpful, and it gives each member the information they need to make their own availability clear. There might be in-laws and aunts who can be helpful but only if asked. Don't be afraid to go beyond the bounds of traditional family lines and ask family friends for help too.

Transparency

Much like an organization with complex abilities to delegate and share tasks, family members are there to help and need similar guiding principles. One such ideal is to have transparency across parties that work together.

In a 2012 article called "Developing Ethical Leadership Skills" posted in Blog.ReadytoManage.com by Dr. Jon Warner, Gael found a synopsis that parallels a way of setting up a "family caregiving system" that might include multiple siblings or an extensive network of those who are stakeholders in family care. Although the content is in the form of business advice, the need for people to effectively work together is not limited to business:

> In the final analysis transparency in business should be seen as a way to keep employees broadly apprised of organizational activity insofar as that activity affects them and their personal interests. It is a way to maintain trustworthy communication between people and senior leaders in an organization that needs these people to give of their best. The key here is

therefore to appreciate how any organizational activity affects different individuals, teams or departments and their interests. Of course, this assumes that an organization is in close contact with its employees and knows their concerns. Few organizations are as close as they should be and may inadvertently keep individuals and whole teams in the dark when they should be aware of what is happening. The first step to being more transparent is therefore to listen to what people need to do their jobs well, and, within the bounds of legality and commercial reasonableness, aim to provide it.

Think of *siblings* in the quote above in place of *employees*. You can see how keeping family members in the dark only postpones problems that can eventually arise, and it often perpetuates them. Having open and forthright communication and sharing routinely may feel like an effort, but doing so will promote family engagement and encourage proactive ideas. Help that may not have arisen if family were not clued in can arrive more easily as a result.

The difference between sibling teams that succeed in a family care system that creates strength as opposed to those that don't can be summed up in the guidelines below. They address upholding trust, having good intentions, talking openly about care, relaying information to all involved, and sharing roles even if it involves disagreements. Reexamine your own ideas about delegation and sharing and see if there are places where you need to be more open.

- Embrace mutual trust and conflict. If the definition of trust is the ability of group members to show their vulnerabilities and frankly state their opinions to their peers without fear of reprisal, never is this more challenging than among family members who have a long history or with doctor relationships that hold a hierarchy of top-down management.

- Team members must be confident in the good intentions of their fellow siblings or family members. Be authentic and open, don't hide information, and be transparent in communications.

- A lack of confidence in the magnanimity of your teammates can lead to defensive behavior and a reluctance to seek assistance or consider the point of view of others.

- Overcome fear of conflict by embracing healthy disagreement. This relies on a level of trust that, once established, recognizes the sincere attempt to push through issues. An emphasis on good communication and respect for individuality is important in healthy disagreements.

- Having a key family contact identified to communicate with your family member's medical team also keeps the problem of too many contact persons to a minimum. If you are the family member assigned to the contact role with a physician during appointments or after them, keep good records, share as needed, and relay any information about the person who knows the patient best: you.

What You (and Your Family Members) Should Know before Doctor Visits

If you are the *primary* caregiver on the front line, knowing and preparing for medical visits with the right information is vital. If you cannot attend appointments, have a backup caregiver ready with the right information and versed in what to expect. The more people who are cross-trained in family care responsibilities, the easier it will be to delegate when they are in town, when you are on vacation, or when you are able to share the role.

For a child, this may mean having a relative who can step in and have the proper permissions and even paperwork allowing their presence in a setting where privacy and other safety concerns govern their care. For care of an adult family member, they can often allow permissions verbally and have the medical staff alerted to their presence and thus prepare HIPPA (Health Insurance Portability and Accountability Act) paperwork if it's not on file. Also know that any key decisions and financial matters related to power of attorney will require only the persons named in those legal documents may need to have a say or be present in matters.

Helpful things to keep in mind if you are managing the physician interactions and follow-up include:

- Items at the ready for hospital or first-time medical visits:

  o insurance card/information
  o photo identification
  o any required copay
  o medicines your family member is on
  o a listing of doctors who might need to know the outcome of this medical visit

- For ongoing medical concerns with follow-up and continued doctor's visits, be ready for this information to be presented, discussed, or inquired about:

  o Any kind of tests wanted or requested (i.e., laboratory and blood tests)

- Questions about procedures that may be recommended (but may not be covered)
- Requests from hospital/lab and current physician
- A copy of any inpatient hospitalization records (if these are not in your possession, the family member or medical power of attorney for the family member can request from the medical records department where hospitalized)
- Keep a notebook as a way to keep a habit of documenting happenings so you can reflect upon them or share them with others. Bring a notebook to every medical appointment and

take careful notes. University of Pennsylvania's Medicine's Cancer Center offers families guidance through cancer care with their site on "Navigating Cancer Care" at www.pennmedicine.org/cancer/navigating-cancer-care. On this site, they advise that those with cancer and their family members should "make the most of the medical team by keeping a journal listing medications, dosages, timing, effects, history, tests procedures, vitamins, supplements, concerns." Use a notebook or other repository to document the medication instructions, names, options, schedules, recommendations, and suggestions—basically anything that seems important and might need review.

- File medical records and access them if you need to in building your case. Ask for copies of all new records too. Keep all notes in one place and use them for the reference that medical personnel often need to evaluate the whole story. Let family members who are open to being part of the care system access things that can help when they need to see the background to maintain openness and transparency.

As the primary caregiver, you are the keeper of the thread of care. Was there a change of mood, appetite, sleep pattern, or symptoms? Are old symptoms no longer there? Recognizing change is a keynote to proper care. You are the one who will notice what others may not, but there may be another day when a person managing the care sees something. Have an open channel to communicate and work to ensure involvement from those who can stay engaged.

The Family System

Your family is connected through care, thereby making all of you part of a family system. Know that this system is not regulated to see each individual, their feelings, frustrations, and all the concurrent needs they may have. It must be communicated, and members of the family system must stay open. For eldercare in which adult children are involved in shared care, familiar dramas from the past may play out again, emotions may run high, and words said that are regretted. Balls may drop—or there may be too many people vying to be in charge. All of these are facets to be aware of within your own family system.

We recommend Francine Russo's *They're Your Parents Too! How Siblings Can Survive Their Parents' Aging without Driving Each Other Crazy* (2010). The author is a journalist who uses her insights to guide adult children of aging parents in how to win crucial help from your siblings instead of letting them drive you toward disarray.

Use the Power of Saying No

As a caregiver, you must know how and when to engage the power of no. Be aware that you can push back and win. When Gael's mom was in her last few months, she became very frail. Until that point, her regular visits to her oncologist were a source of care and camaraderie since she had known him for many years. It felt personal. Inevitably, the trip became more burden than care. It took a tremendous amount of prep time to get her from bed to car to walker to wheelchair and back again as every trip became more dangerous with weather and curbs. What the physician actually needed was her blood work rather than her physical presence, and Gael suggested "no more office visits." It was easily arranged for blood to be drawn and sent to the office where the numbers could be monitored from afar, and since the doctor agreed, it was a tremendous relief for all concerned.

Gael also exercised the power of no when her son was recommended as an "ideal patient to receive an experimental pharmacological cocktail" to establish normal circadian rhythms while he was struggling out of unconsciousness following his traumatic accident:

> I was given paperwork to review and sign off on, yet
> I noticed the demeanor of the male nurse who was

shaking his head "no" while I was reading the papers. It took that shake for me to realize perhaps this was a choice I had in front of me, not a given to sign off on. I began to question the reasoning for the apparent need and realized right there and then that there are indeed situations where caregiver choice is present yet not offered, and this was one. In my son's case, he was in an interior room with no natural light, so I asked for him to be moved to a room with a view before testing out a drug intervention. It worked. Within twenty-four hours, he opened his eyes.

The ability to say no or trust your gut are vital elements of being a family caregiver, and as an advocate and primary support of another person, they are important and must be used. We who are not trained physicians cannot hope to have the kind of knowledge and experience that is born of profession and education, yet many situations boil down to the common sense of the caregiver. Exercise yours and share the insights you gain with others helping you in care.

Chapter 7 Reflections: Acknowledging Our Roles

- Can you identify where you find the most challenges with boundaries? Why?
- What next step are you preparing for? Be ready for events that you know will be stressful by finding out information early. Keep records and jot down notes.
- What feels difficult for you to say no to? Monitor your stress level and be ready to admit to yourself what you are afraid to say no to. There are stresses that are meaningful, stimulating, and just plain hard. Put them into context but also be ready to put into place boundaries for the future that you will share with others.
- What of probably events can you jot down now? Think of upcoming events and anticipate possible outcomes and how you will handle them. Be ready for events to happen—and take preventative actions to reduce concerns.

Notes to Myself

CHAPTER 8
Care and Work

Making it easier for more Americans to be the workers and family
members they want to be will make our economy and country
stronger. Companies that stand by the people who work for
them do the right thing and the smart thing—it helps them serve
their mission, live their values, and improve their bottom line
by increasing the loyalty and performance of their workforce.
—Sheryl Sandberg

It's Adelaide: Gael

*Cousin Adelaide held a place in our family more akin to the younger
of my father's siblings than the niece she actually was to him. She was
called Little Adelaide even though she was measurably taller than my
father's sister, her mother Big Adelaide, or "Big" for short. We often
heard stories of their upbringing, like the one where the siblings all
had to pile up to sleep in the same cold room while growing up during
the Great Depression, with Little Adelaide right there in the heap.
She had a devotion to family—her uncles in particular, I think. A*

photo of Nana, whom she adored, was displayed among her few visible mementos last I visited her in hospice.

In the early days, she was a frequent visitor to our Brooklyn apartment. As a girl, I enjoyed her captive audience to my song-and-dance routines—a study in patience I am sure—and she showed no compunction evening the score by beating me ruthlessly in board games, out to win no matter my tender age. I always knew she was feisty—a real competitor.

Adelaide had a distinctive, gravelly voice, and an unmistakable, if somewhat sarcastic laugh. Everyone who knew her could mimic that voice at the other side of the phone with hardly a skipped beat. "Hi, it's Adelaide," she would say as if we didn't know. The gravel deepened over time—as did her ferocious dedication to smoking.

Her insinuation into our lives was somewhat of an annoyance at times, and with no children of her own, that competitive spirit became the highlight of her career woman status.

I recently had a memory of Adelaide, with her freckly skin and shock of red hair, joining me and my two boys at the beach for one of our daily summer sojourns. I was just a young mom, the sand and sun a welcome treat, wondering if Adelaide might actually need to be swathed from head to foot in flowing white like some escapee from the set of Lawrence of Arabia. I nonetheless relaxed on the sand in the tiniest bikini I could reasonably get away with. Those were the days, and times do change. My favorite accompaniment now is a large linen scarf draped over my shoulders to go with my SPF 50 and a beach tent. Adelaide was a woman ahead of her time, and today my ever-present linen scarf is a poignant reminder of having roots, and the different kinds of care we give to ourselves and each other.

Now all of the siblings are gone, and the cousins have moved away. When it was Adelaide's own turn to die, she seemed a lonely one indeed but for a handful of stragglers from the neighborhood and her workplace. Somehow, she had managed to elicit a sort of familial devotion from a makeshift tribe that was nothing short of remarkable given her erratic temperament. They were few in number, but they were loyal to the end. In the final analysis, an unexpected clan had arisen *out of everyday life and care, a clan that expanded the hearts of all concerned, including my own.*

Care and Work

Work creates a home away from home. Work is a place where a role is expected, and until caregiving roles become too much, work duties often do not overlap with home care duties. Until they do. Unexpected or prolonged care of another creates a creeping effect, and the need to explain "what is going on at home" is nearly unavoidable when working with others who are wondering what has changed in the person they know but whose home life responsibilities have changed. Working and caring for others can become a conundrum for some. As this chapter will elucidate, when it is openly addressed and a strategy is devised to continue working amidst difficult care, it can be managed. And the tribe of coworkers and employment can act as a stabilizing factor in seemingly unmoored situations at home due to family care.

Managing care of family with work can be a delicate balance. Family emergencies, such as a parent's heart attack or a late-stage cancer diagnosis, require swift action. A traffic accident of a spouse or a broken hip of a great-aunt with no children requires a long recovery and possibly a new normal with the need to triage events at home and little time to plan for how work can still be done with as little disruption as possible.

Regardless of what the situation is at home with your family member's diagnosis, you will have to prepare for changes to your work routine to minimize affecting work in progress. Life might feel hurried and forced as you add doctor's visits to your lunch hour or place phone calls with hired agency's caregivers to act in your place when you can't be at home. Managing a continuation of work will take conscious effort. There can be a sense of grappling through the day with a tug back and forth. Decisions and priorities will now be more complex for your family member, and you will have to be ready to explain the situation to your coworkers and bosses. Your actions may become layered with guilt, and you may waver about how much to divulge to maintain the privacy of the person you are caring for. Management may be concerned about you maintaining your workload.

A strong sense of knowing what you are doing is right with respect to your care will allow you to feel grounded in this harrowing time. As you ask for accommodations with your employer, you will have to be resourceful. There may be times when you will need to advocate for flexibility to do your work in a different way than before, and there may also be times in which you may ask to work less or not at all.

Overall, it will be vital as you navigate through the challenges of family to be strategic about your work priorities and remain open to a flexible work schedule. Try to have an understanding of how others in similar experiences fared during short- or long-term family health crises. The Human Resources (HR) department is a vital stop to find out about options with respect to working during a time of exceptional care responsibilities.

A Strategy to Manage a Busy Life

You may not be able to take on all of the care at home and perform as you had in the past. To work and perform the role of caregiver, you will need a strategy to keep work and home life managed, and that could involve hiring others. Begin by trying to learn how commitments at work can be met and your caregiving goals achieved without a loss to quality of care. If the care is for an adult, decide where you can delegate to family members or find an in-home care service. If your family member has advanced dementia, adult day programs exist during work hours. They may be able to provide care coverage during your workday. If you use a paid company for in-home caregiving services, research their background, and if you feel more vetting is needed, look to companies like Care.com for background, bonding, and other licensing proficiencies that might be needed for care roles you can't take on. You may need to evaluate a facility in which to place a family member. This process can take time and require buy in from the person cared for as well as family members.

Financial implications will be important in planning for housing. Rising costs will be a factor, and insurance may or may not be able to mitigate the impact of the family contribution toward care. Having paperwork for key documents with respect to financial and medical power of attorney documentation will be essential in the care-implementation plan you will carry out.

Managing Expectations at Work

With respect to work, be ready to have an open dialogue with management if necessary. Be frank so you can manage expectations of your commitments. Also be prepared to indicate whether other staff will need to help cover your workload to meet deadlines. Some employers have neither the staff nor the budget to do without your presence for work commitments, and they may need to reallocate existing staff or hire interim staff. Try to creatively evaluate how you can meet work needs and be away during agreed-upon times such as lunch hour or after work.

The perceptions of others on the job will be part of your thought process as you address caregiving commitments. In the workplace, people may not know how taxing this type of family care can be. Even if you are working at home, your absence will be noticed. If you begin to miss deadlines or are not as engaged as you had been before, you'll need to be able to have the conversation about what is happening in your home life. Be aware of these challenges and prepare accordingly:

- Be proactive with your manager and the Human Resources department to explain situations and their duration and negotiate where necessary. They are experts in areas like taking leave using vacation, or taking advantage of other benefits within your company. They may offer for you to have flexible working arrangements as long as your supervisor is in approval. None of this can be done if you keep your caregiving a secret or don't go through the formal channels to bring it to light.

- Be aware of defending absences or hours away to those who may not be in the know about plans you have developed with a boss or management.

- Be ready to meet commitments with work done outside of traditional hours or have delegated to others (all the while, plan to accurately show how you met commitments you are working on).

- Some people will not understand (and that's okay) or will see your caregiving as an excuse for low performance. Maintain awareness of these perceptions—but don't let them get in the way of your convictions to provide the care in the manner you feel is best.

- Use discretion with what you do share and how you divulge your own ways to manage caregiving (and the life and experiences of the person you are caring for),

- If you feel that exiting the workplace is important, see if you can take a leave of absence or go part-time rather than quitting. Maintaining your workplace identity is an important matter to evaluate, and leaving work altogether may have implications you haven't given thought to in your complicated life.

- Research work-life balance and learn about others' experiences in managing the stressors of care and staying engaged in work—while not sacrificing home commitments or obligations to family members.

Using Workplace Programs to Your Advantage

In larger companies, programs at the workplace level are working to ease caregivers' lives. Employee-Assistance Programs (EAP) can help caregivers get free counseling and advice from experts. These voluntary programs are available to employees. EAP programs offer access to resources or confidential assessments to help employees find solutions that help them balance their lives. There may be employees who wish to seek short-term counseling, referrals for child or eldercare services, or follow-up services to aid with personal and/or work-related problems. EAP programs tend to be helpful for those who are seeking help for issues affecting their mental and emotional well-being, which can often happen to those who are caregivers. EAP counselors also work in consultative roles with Human Resources staff, managers, and supervisors who can address employee challenges from an outside perspective. Outside referrals are made through a confidential avenue and are helpful for connecting caregivers to resources in their communities—sometimes at no cost.

Approaching Human Resources in your company can provide insight to a company's corporate abilities to help with work-life balance strategies or offer advice if federally mandated options are appropriate. The Family and Medical Leave Act of 1993 (FMLA) requires covered employers to provide employees with job-protected and unpaid leave for qualified medical and family reasons. FMLA leave can provide up to twelve weeks of unpaid, job-protected leave per year. Companies that are large enough to have FMLA must provide group health benefits that are maintained during the leave as long as this time away is to care for an immediate family member (spouse, child, or parent) with a serious health condition.

As employees around the nation struggle with family care, the Human Resources community is trying to adapt. The company of the future will be changing and adapting its policies and may be willing to hear ways you can adapt and do your job from home or during alternative hours—or even help you arrange to take paid time off or vacation to support your care needs.

Supporting Working Caregivers on a Federal and State Level

Other first world countries have different views on family care and whether it should warrant time away from work (with or without compensation) when workers need time away to manage care. America's social fabric has a safety net in terms of disability and social services for those who meet specific criteria—through the Family and Medical Leave Act—but many smaller businesses do not have to follow the law due to the limited number of employees on their payroll.

Prolonged family care can be unpredictable, and it can be hard to have a consistent federal or state policy align within the twelve weeks. In addition, for many, having no income for twelve weeks can be financially detrimental in a stressful time. In addition, family matters who could be covered resourcefully in one family and be lacking in another create problems for policy developers.

John McKnight, an author and policy critic, addressed this in *The Careless Society: Community and Its Counterfeits* (1996):

> Since love is not a political issue, care is not a policy question and service becomes the one business that is an unlimited, unquestionable, and nonpolitical "good."

McKnight shows how communities rely upon medicine and human services for support, but these professional areas often can't help. When individuals try to find answers for themselves, including finding unique individual workplace solutions, that path may provide better results than a federal or state policy enacted to help large numbers of people with different needs.

State programs are also available, but they can vary widely. Caregiver.org provides a fifty-state caregiver profile of each state's community-based services and noted variations in states. For example, one state might provide caregiver respite programs or home / community-based service waivers for caregivers, and others do not.

Respite Care

For some, having a break from care in terms of time away from a family member is important, especially if work requires you to be away on a trip when you need coverage for a family member in a facility. *Respite care* is a term that implies rest and is largely used in the context of giving caregivers a break from family care by removing the patient from the home so that the caregiver can recover. For the millions of Americans caring for an aging or disabled loved one, respite care can serve a number of valuable functions. Respite is often used when:

- the family caregiver needs to travel
- the family caregiver needs a break
- a family caregiving recipient needs a temporary change of pace or a break
- a family is gradually easing their loved one into life at a senior community

Because of the high stress associated with placing someone you care for in a facility, allow yourself time and space to manage these steps. Your family member will have a range of negative emotions associated with the change in residence—even for a short respite stay. Although it may create a sense of guilt or anger from the person entering a facility for a short duration, it will allow family caregivers an opportunity to recharge their batteries. It's important to keep in mind that the cost of respite care varies with the type of agency and services the person needs, and this must be evaluated before approaching a respite program. Fortunately, financial programs may help pay for respite care. For example, hospice, long-term care insurance policies, state programs, and other benefits may cover some of the cost of respite care.

Solopreneurs

The self-employed often have no employee-assistance type programs that offer caregiver resources or FMLA (Family Medical Leave Act) benefits like entitled leave to fall back on at all. The most significant thing you can know is that the care you give is a choice—even when you feel there is no other way. Choice itself gives you the power to enroll whatever resources do come your way. Ask for and say yes to receiving help. Stay open to areas of support you might not expect.

Gael was called across the country for her son on a sunny Sunday after having spoken to him just two hours before. As an expert skier, he was overjoyed about opening day of the ski season. First day, first run, unexpected outcome. As a wellness studio owner with a full schedule of classes and clients at the time, Gael had no preparation for leaving her entire business in someone else's hands on a Monday morning. Yet leave it she did when she took the call that announced, "Your son has had an accident. Pack a bag, get on a plane and be prepared to stay". Student teachers organized themselves and stepped forward to cover all the classes. Keys were made and distributed as needed. Someone picked up the mail. Someone else took on communications. A spiritual community she mentored took up a spontaneous collection and sent her a check! Unplanned and unrehearsed, it all worked, but flexibility and a sense of allowance were vital. So was good communication and delegation.

You may have to change your work environment to accommodate your higher goal of caregiving. You may even have to give up your work. You will certainly lose income if you are a solopreneur, and you will miss opportunities. You might not always feel you have a choice, but how you approach caregiving is always a choice. As long as you remain in the knowledge of that aspect of your responsiveness, you will manage. Let others help you.

Investing in Support for Caregivers

Although you may expect support in your role as a caregiver from coworkers, it may not always materialize. Expect it anyway. Remember that work-life balance is a discussion happening in our society already. Coworkers may learn from your experiences when they face similar matters.

Experts agree that if we don't invest in supporting our millions of unpaid family caregivers, we will pay to both support those who are cared for and those caregivers who are burned out. Companies know this is an investment that is worthwhile because helping their employees shows they are valued in return. Providing benefits related to family care can reduce absenteeism and turnover and improve organizational health and talent acquisition. However, not all caregivers' companies have far-reaching benefits that help them.

Jennifer in Sharon's 2013 dissertation on family care in *Bridging Generations: Personal and Societal Implications from a Study of Sandwich Generation Caregivers* made a comment that is a widely held opinion how they would love to see government funding for something like the support of caregivers. There is an understanding among those who are caregivers that society needs to make more of an investment in this "army of unpaid help." The future of census trends tells us that employees will be caring for families at greater rates with an aging elder population. With a shortage of public policies that help workers, the workplace will need to better meet the needs of their staff. With company trends placing an importance on helping their working caregivers, it's clear that investing in talent also means investing in the families of the valuable employee they have now and will recruit in the future.

Bold new steps are being taken as employers recognize the needs of workers as they move through different stages in life and have family in their care whose needs shift. With a typical company having a sizable number of employees, providing flexible family leave to support a range of life events that impact them and their families—from pregnancy to surgery to family care of another person whose health requires the employee to be with them—the notion of flexible time off is a valuable company perk. However, for those who need more time away or who can work and care without time away, many companies are creating policies that embrace intermittent care and allow workers to take advantage of flexible hours, telecommuting, or job sharing.

Technology companies in Silicon Valley have taken the helm in ideas like "unlimited time off" for care and are testing the waters for others who are watching to see if these ideas will be followed or if constraints will ensue. Yet unlimited paid time off (PTO) with sick leave and personal days is rare beyond the biggest companies offering this novel approach. A *Society for Human Resource Management* (*SHRM*) article by Joanne Sammer (2014), "Unlimited Paid Time Off: A Good or Bad Idea," evaluates data showing that less than 1 percent of employers offer such a unique policy. "This is anything but an emerging trend," said Bruce Elliott, *SHRM*'s manager of compensation and benefits. Although it would mean that Human Resource staff would no longer have to track this type of administrative time, it could allow staff in the office to feel a sense of unfairness or as though some are working harder than others who use this benefit less.

Time will tell if companies adopt this practice for employees. For some who work in these types of Silicon Valley companies offering unlimited time off, many feel that taking "too many" days could get in the way of future promotions. Unspoken rules and perceptions may trump written policies. Thus the jury is still out about whether this approach would actually succeed on a large scale.

Hired Care

Not every hired care employee is especially well suited to the job or has chemistry with your family member. Be aware. Don't be afraid to ask questions or look for another company or person. Know the details of agencies that provide care at different levels of expertise and how the pay varies. More experience with medical procedures, nontraditional working hours, holidays, and late nights will cost more to hire. Some companies are bonded and can have an assurance of coverage with a network of caregivers. They might also have hourly minimums for which you will need to be prepared to have the hired help be of use or standing by to help. It is possible to find other hired help—not found through a company—with more flexible terms. With any outside help, valuables and confidential things should be stored to ensure peace of mind and security. A family member in need may be vulnerable with health matters—so keep a heightened sense of guardedness surrounding the care you are delegating.

Using a Paid Advocate to Help with Care

Hiring a paid advocate to manage care can help with meeting work commitments and life. They can be hired for a consultative meeting or for ongoing management of care over time—all for an hourly rate. For example, a professional geriatric care manager is a person who has been educated in human services—social work, psychology, nursing, gerontology—and with this training, they can help plan, coordinate, monitor, and provide services for the person you are caring for. Advocacy is the primary function of the care manager, and they often belong to an association with standards they follow.

Working with an outside expert offers support and can help you create a care plan that allows you to work and care. Ideas to balance work and care roles might mean communication strategies to engage other family members, employing a caregiver for shifts that family can't cover, or even shifting career plans to accommodate care and work in a different sphere.

Family care affects all ages of employees and deserves credence it often doesn't have right now. Until the future when care is addressed in our organizations in a beneficial and useful way that helps family caregivers, it is up to us to explore, evaluate, and, as warranted, adopt and share ideas that work in our times of need. If we engage geriatric care managers, work with employee-assistance programs, or evaluate the next steps on our own, the journey through caring while working will require patience, persistence, and a commitment to balance.

A wide range of personal consultants have felt the need to give back after their own experiences. In addition to being willing and able to help you through the caregiving years, Certified Caregiving Consultants are family caregivers or former family caregivers who can guide you through your work/care situation. Gael joined this cohort of trained experts as a Consultant after her own experience ignited her desire to shift the emphasis of her business. With a range of skills, backgrounds, and resources, consultants understand where you've been and where you're going. You can find a directory at www. CareGiving.com, join one of the chat rooms for immediate support, or look at the corporate-solutions page for ongoing workplace initiatives.

The comeback
is always
stronger than
the setback

Chapter 8 Reflections: Care and Work

- Recognize how your contributions at work and home are in sync or out of balance.
- Make a list of steps to minimize the stress at home or at work.
- Think through how your life as a caregiver could improve with some proactive actions on your part. Use the following prompts to guide your thinking based upon your situation:

  o If you work for a company that has HR, find out how they help caregivers. Think about ways you may be holding back things at work that, if others knew, might help you without disclosing too much. If people are not aware, they can't take steps to aid in your journey. See if there is a caregiver-support group at work or if one can be set up to help employees in your situation.

  o If you are a solopreneur or work for yourself, you will have to think outside the box. List people who can take over some of the work in whatever way they can. Ask. You might also search for a virtual support group or organization in your community.

  o If you need to evaluate cutting back your hours or taking leave (short-term leave might mean using FMLA, and long-term leave might mean leaving your job), reflect on the effects of this shift in your life and how it will affect your financial situation.

- Scour your thoughts for preconceptions that might be limiting your potential for receiving help from unexpected sources.

Notes to Myself

CHAPTER 9
The Best for All Concerned

The number one thing caregivers can do for other
caregivers is to say, "You are not alone."

I Spy: Gael

*I've enrolled in a writing class for memoir. The effect is surprising as I
observe myself reliving events, going this way and that, all the way back
to childhood, even while I'm asleep. My dreams are tied up in memories,
the writing bringing them forth both in and out of consciousness. I can
choose any topic to write on, but I find myself immersed in family life
and touching deeply with death. Today, I remember my mom playing
"I Spy" from her wheelchair while spotting geese and other wildlife
during our nature outings. This occurred during the last year or so of
her life, a rather dramatic shift in what she liked to focus on at all, and
the memoir I've decided to write is making the memory so clear.*

The road between us wasn't always easy to understand. This was partly due to her fear of my rebellious generation, a fact I could do nothing about, but when I later became an outdoor guide traveling for weeks at a time, she could certainly see my absence but not my joy. Without knowing the majesty of mountains and open skies, my time spent in distant places must have seemed a dismissal of all she held dear. In essence, slowly but surely, I was digging a relationship grave that I wouldn't get out of for a very long time. The funny thing was that I didn't really see how deep it was when smiles covered the distance as surely as the miles.

Back to "I Spy".

By the time my dad got so sick that the entire family needed to rally, I had the opportunity to move from New York to Pennsylvania where they lived. Now we actually were close, at least in proximity. I handled my dad's illness along with his growing needs with a sort of pragmatic patience, but when my mom started winding down, it broke my heart. Her growing forgetfulness seemed to have a reverse effect on her "safe-at-all-costs" demeanor. Suddenly she was waving at strangers in the lobby of her care facility and participating in a painting class with unknown residents. She blew kisses at the nurses' aides and sometimes kissed their hands when they were especially kind. She laughed a lot, and she submitted to my insistence that our visits would now be "outdoor adventures." My high-peak climbs moved to the nature sanctuary, and turtles became our own version of moose and black bear. We learned a lot about each other on those outdoor trips, and I got to witness her growing excitement that nature was our friend.

I still frequent the Black Rock Sanctuary on my own, and I was musing about how I might add an "I Spy" photo to that memoir and capture the experience through my mother's eyes. At that very moment, while walking the well-worn path, I came around a bend—and there it was. A single goose paused at the water's edge, just waiting to be seen.

We never know when our biggest opportunities will arrive, when years of distance, or efforts lost, will suddenly open to become a time of recognition and change.

One evening before she died, she leaned in to me in a moment of crystal clarity, and said "You know, I never understood you." I said, "I know, Mom," spoken from a deep well of understanding and love. It was enough, and hugging her, I tucked her in.

A Growing Number of Caregivers in Need

There are many caregivers vying for support, and their circumstances vary. In September 2016, the National Academies of Sciences, Engineering, and Medicine issued a report titled, "Families Caring for an Aging America." The report stated that at least 17.7 million individuals in the United States are providing care and support specifically to an older parent, spouse, friend, or neighbor who needs help because of a limitation in their physical, mental, or cognitive functioning.

The circumstances of individual caregivers are extremely varied. They may live near or far and provide varying lengths of care. The caregiver may help with household tasks or self-care activities, such as getting in and out of bed, bathing, dressing, eating, or toileting, or may provide complex medical care tasks, such as managing medications and giving injections. The older adult may have dementia and require a caregiver's constant supervision, and the caregiver may be responsible for all of these activities.

Care of elders takes center stage as this demographic of the population is growing at a fast rate. With support from fifteen sponsors, the National Academies of Sciences, Engineering, and Medicine focused on the topic of our nation's family caregivers of older adults and came to recommend policies to address their needs. This involved examining how to minimize barriers these caregivers encounter in acting on behalf of older adult family members "Families Caring for an Aging America" (2016) provides an overview of topics including personal impact on caregivers' health, economic security, and overall well-being. A PDF of this report is available at The National Academies Press website (nap.edu) to find the full report sold as an e-book or in hard copy form. Alternatively, it can be read in a brief summarized format at the National Academies Press' website at no cost.

Palliative Care and Hospice

When we position family caregivers as leaders who have a vital contribution to make, we often see our fierce and protective nature step forward most clearly as end of life issues draw near.

Dr. Ira Byock, author of *The Best Care Possible* (2012), is an international leader in end of life care and he speaks of dying well. From his medical career spanning decades, he knows the challenges of caring for those who are advanced in age or are chronically ill or worsening into a terminal condition. He emphasizes palliative care, which is useful at any stage of a chronic or terminal illness and physicians feel is best provided from the point of learning of the medical diagnosis. Palliative care focuses upon relief from the symptoms and stress of a serious illness. Quality of life is central to the goals for both the patient and the family. Palliative care is also important for helping patients to understand medical choices in directing their treatment. It should be known that palliative care can be provided alongside of curative treatment.

In palliative care, you do not have to give up treatment that might cure a serious illness. Over time, if the doctor or the palliative care team believes ongoing treatment is no longer helping, hospice can provide comprehensive comfort care as well as support for the family, but, in hospice, attempts to cure the person's illness are stopped (with more details available from the physician as to what this means for the patient specifically). This does not mean that if a person has cancer and they develop an illness, they can't have treatment. Additionally, the palliative care team could continue to help with increasing emphasis on comfort care with the goal to prevent or relieve suffering as much as possible and to improve quality of life while respecting the dying person's wishes.

If you are choosing to use a Hospice, you can research organizations in your area and find different non-profits or for-profit organizations that have vary approaches too their care. There is an inter-disciplinary team of professionals which not only include nurses, but also social workers, spiritual care coordinators, and even volunteers who attend to the needs of the dying and their comfort and make the experience manageable for all.

It is true that medical professionals sometimes treat these symptoms aggressively with varying results and that often less aggressive treatments or care plans can impart a better quality of life. There are many instances in which treatments that are too aggressive and invasive leave the patient tired, worn thin of hope (and this can leave with caregivers to manage the emotions and treatment options available to the loved one). Dr. Byock states, "Caring for a loved one who is in the waning phases of life is inherently hard. All of us will eventually face the end of life, but neither dying nor caregiving has to be as hard as they are today." Exploring options with care and how these treatment plans may affect quality of life could be a valuable use of time and will leave open choices to be made instead of assuming only one course of action.

Secondary Prevention

Dr. Byock is a proponent of the notion called *Secondary Prevention*, which can help prevent crises and stress-related illnesses among family caregivers by having health-care professionals look out for the health of caregivers as they are vital to the framework of those they care for. Secondary prevention can be taught to physicians and others in the medical community, but it requires training. Many medical professionals are already managing patient loads and taxing work schedules, and training may not allow for questions to caregivers that would also help them see signs of burnout or health decline from exhaustion. Those who are the primary concern are already cared for through the medical complex, but there needs to be ways in which those who are part of the web of family support are on someone's radar, especially when fatigue and exhaustion cause physical or mental ailments.

Sharon had a participant in her caregiving study "Bridging Generations" (2013) for whom the idea of secondary prevention would have prevented a burnout experience. Paula was taking her young daughter to multiple appointments for a neurological condition that was not fully known or understood, and the symptoms were perplexing to the family. Paula didn't notice the toll it took on her health, and her daughter's physician never asked her about it in the many visits to the medical facility:

> Shortly after my daughter's health diagnosis, I was at work and basically collapsed. I was taken to the closest ER. I went back to work three days later and was driving home and felt the same symptoms and was later taken again to the ER. This time I was admitted and had a stay of six days due to exhaustion.

Paula had to learn to take control of her health in a way that meant not sacrificing rest, having good boundaries with scheduling, and making sure her health did not worsen. These skills have been helpful, and Paula has since taken care of other elders in her family through to their passing—without needing to be hospitalized.

In 2018, the American Medical Association released new guidelines for physicians designed to provide guidance on caring for caregivers who often arrive at appointments and treatments with their loved ones. The information for doctors is found at www.ama-assn.org/sites/defult/files/media-browser/public/public-health/caregiver-burnout-guide.pdf, and is titled "Caring for the Caregiver: A Guide for Physicians." In this downloadable brochure, the AMA explains caregiver burnout and how this can appear in the form of role confusion, unrealistic expectations, and even asking the physician for unreasonable demands.

Although the physician isn't treating the caregiver during the family member's appointment time, resources can be given to the caregiver if burnout and decline appear in the caregiver's health. With caregivers speaking up for what they need we can expect the medical community to adapt practices that support caregivers. In another medical context, after new mothers give birth, they often take questionnaires on their medical and emotional health which are followed closely by their pediatricians because the well-being of the mother is linked directly with the health of the child. The doctors and medical staff who look out for signs of postpartum depression, and other medical areas should have similar guidelines in place. The best skills caregivers can adopt is being their own advocates in advancing and looking out for themselves while caring for others and reporting any symptoms to their own doctors.

An Urgent Need for Action

Society needs person-centered health care, and it needs to focus on family-centered care. Below are some ideas that may help caregivers if federal, state, and local policy makers and organizations aim to help those in the family caregiver demographic:

- Create an approach to addressing caregivers as a demographic and support the essential role of family caregivers to older adults. This strategy could look at how communities and workplaces might support their health, values, and social and economic well-being. This is so important as our society grows older in larger numbers, and it is culturally and ethnically more diverse than ever before.
- Evaluate how Medicare, Medicaid, and the US Department of Veterans Affairs might improve services for family caregivers because they are routinely called upon as patient advocates. Caregivers are on the front lines of health care service use by family members in health-care delivery, and soliciting their

input with payment and models of care may help with service delivery and payment collection.

- Family caregivers should be able to access available support to help with care, perhaps by using online systems of dissemination that are accessible in every state. Caregivers should be made aware of home- and community-based programs and services they can use to their family member's benefit whether it be adult day care services in a community or state-funded respite care they can take advantage of.

- Expanding the funding for programs that provide explicit supportive services for family caregivers like the National Family Caregiver Support Program are also a vital way to ensure that evidenced-based caregiver intervention programs reach more people who need them. Although the Administration for Community Living began providing grants to states to cover programs that support caregivers meeting service needs intended to reduce caregiver stress, grants like these need further funding to expand their reach.

- Innovation will help working caregivers which is a fundamental need. Our leaders should explore, evaluate, and share best practices that support working caregivers to improve benefits for those who work and allow flexible work accommodations—and even paid time off in a consistent and practical way—for those who need it to help family members.

- Improve upon and expand the data-collection infrastructures by working within governmental entities such as the Departments of Health and Human Services, Labor, and Veterans Affairs to analyze and understand how to serve these populations and their families. This will work to improve tracking and reporting on the experience of family caregivers and allow for outreach to benefit those who need services at their time of greatest need.

- Create commonality with care programs executed across state governments to aim for more consistency. This allows families to have equal chances to overcome challenges of care with equivalent supports and equity from programs that are similar from state to state.

Until these actions are put into place, it is the job of the engaged caregiver to find your own answers and solutions. Try looking to organizations such as AARP and Family Caregivers Alliance for answers—or write to your state/congressional representative to ask for progress on the above types of initiatives. Elected officials offer ways to connect including online access for easy letter writing.

Four-Direction Framework for Balance

You might find it useful to engage the following concepts when you feel yourself faltering. The four directions are a balanced framework for harnessing clarity, integrating courage with compassion, embodying wisdom, and engaging in the practice of letting go. Like a wheel, they will help you progress with the greatest ease possible.

Remember that expression about how the parent (or caregiver) should apply their oxygen mask first before the one they care for? It is wise advice and a helpful reminder that you need to care for yourself so you can take care of another well. Whatever the outcome for the person you care for, you will need to take your own place in the world.

Clarity

- Know what you are prepared to do and what you must delegate.
- Every event begins with awareness. Simply be aware.
- Prepare now for the understanding that things change.
- Begin a journal. What are your feelings? What would be your best possible result?
- Access your true intentions regarding life and death and keep them in mind.
- Practice slow and steady breathing, particularly on the inhale.

Courage and Compassion

- Accept the possibility that you may need to adjust your timeline for life.
- Exercise your power of choice.
- Negotiate. If you work, you may need to ask for that raise or push a new project to next year so you can eliminate conflicting desires from your caregiving choices.
- Commit to the critical importance of your self-care in word and deed.
- Park in the "lucky space," which is the furthest spot from your destination that you can manage and use the time for self-reflection rather than dashing this way and that.
- Be grateful.

Embodiment

- Honor your place in the circle of care and recognize that you are among the experts as an advocate.
- Tell your coworkers, management team, family members, and others what you are handling.
- Value transparency in communication.
- Be team oriented.
- Ask for help.
- Stay informed.

Letting Go

- Practice whatever helps you discharge energy—meditation, exercise, etc.
- Continue to relax into specific rituals and habits that support you.
- Learn to say no.
- Surrender to all of this as a work in progress.

- Be aware of and tame the need to control.
- Breathe some more, focusing this time on the exhale.

Finding Helpful Words During Caregiving

In his book *The Four Things That Matter Most: A Book About Living* (2004), Dr. Byock speaks to the qualities of relationships that matter most as we face end of life, and offers stories from patients and families overcoming profound challenges by offering four simple phrases —"Please forgive me," "I forgive you," "Thank you," and "I love you". He reminds us of the enormous power we have to mend relationships and nurture our inner lives. Too often we assume that the people we love really know that we love them, and Dr. Byock demonstrates the value of "stating the obvious" while providing practical insights into the benefits of letting go of old grudges and toxic emotions. His stories help us to forgive, appreciate, love, and celebrate one another and live life more fully.

Chapter 9: Reflections: The Best for All Concerned

- What can you do to say yes to yourself? Can you manage the difficulty of the moment while not diminishing the stressful events?
- Revisit what you can do today to fuel your initiative and have a voice.
- Remember that sharing is caring. Think about what you can do for a caregiver who helps you or a friend who is a caregiver. Can you share in something good? The circle of care begins with you,
- How can you be better at coping? Can you improve in these areas for yourself and the person you are caring for by having a more flexible/expansive sense of self and your capabilities?

Notes to Myself

CHAPTER 10
The Power of Resilience

It is not how much you do, but how much love you put in the doing.
—Mother Teresa

Beginnings: Sharon

I am three months pregnant and so excited to have my second child in six months. My first-born child, a daughter, can't wait to have a sibling as she suggests names for a brother or sister while helping me shop for her kindergarten wardrobe. Moments like these on the weekend alone with my daughter allow us to talk about my growing tummy—only to be upset when morning sickness arrives. The sickness comes and then comes some more. I lose so much weight the doctor has to prescribe me medicine to stop throwing up. Just as I get used to heaving with regularity, I start to feel terror that something is wrong.

One day at work at my job at a hospice, I get a phone call from my doctor's office asking me to find a place quiet where she can tell me about my first trimester screening results. I make my way to the hallway where she tells me I have a high chance of a baby having possible genetic defects and asks me to come for an evaluation to discuss the initial test results. It is a scary few weeks until the appointment, and the subsequent genetic tests to tell us more. The mental ambiguity is further punctuated by my bouts of momentary hyperemesis, making my life feel physically untethered too. But then we get the phone call on December 24 that my second child is genetically "typical," and all is okay. The initial tests were a false alarm. The person asks, "Do you want to know the gender?" After all this, I exhale and breathe the words, "Yes." I learn I am having another girl.

Two years pass, and another pregnancy comes—this time with ease. In the fall, our third child joins the clan. Now we have a son and a new boisterousness as he progresses from baby to toddler. I miss him as I arrive to pick him up with his two sisters. Today is different from other pick-up days. The preschool's assistant director nervously pulls me aside. "I don't know if you noticed this, but frequently at school, when we call his name, your son doesn't respond. You might want to have his pediatrician look into it." I know she sees my expression change as we both read into possible conditions or problems he could have. It's a first of the clues we have laid out before us as parents that something might be amiss with his development. Yet, I brush it aside after my husband doesn't worry because we don't seem to notice the same matter at home.

As my son enters the toddler years, his speech is not progressing. His anger, however, intensifies into bouts of frustration and hitting when he isn't understood. Although we try to avoid reading too much into his behavior, episodes worsen—and he bites and hits more often. In one instance, on an airplane, airline staff had my son and me pulled aside. They were considering not having us fly the next leg of the flight due to his disturbing tantrums on the prior flight. We manage to take the plane, but I am fearful and nervous in every public setting.

We decide to tell the pediatrician, and she refers us to a developmental pediatrician. Our hearts sink. After visiting a doctor in another county and a hearing expert who rules out hearing loss, we learn that we qualify for early child developmental services outside of the medical clinic for speech and sensory remediation. There are more experts to see for my son's care—each one involving a fight in the car over being buckled in and usually a fearful bathroom visit involving the toilet flushing sound. Once he bites me so hard when I buckle him that my upper arm turns purple. I need to take antibiotics. I keep this in and tell few people I know about his behaviors or his services.

As more than a year passes, my son is able to benefit from our local Regional Center's interventional services (that go up to age 3). My son eventually adds words to his vocabulary but his anger and frustration continue to be worrisome. His reactions to sensory stimuli normalize as time passes and we can make it through a typical public bathroom visit without fear of a toilet flush or air dryer setting him off. Slowly as he matures through toddler years, his speech catches up to what is considered normal and no longer needs interventional services any more. His behaviors become more manageable and less impulsive and problematic. My husband and I reflect on how we have developed a manner of parenting that has shifted from the way in which we raised our other kids with more structure, patience and ability to be flexible. As we grow as parents, we know we can manage whatever comes.

The services my son once had for a short period of time now have given me a glimpse into what parents of special needs children go through daily. I see how their lives will be supported by services and providers that they will rely upon far into the future.

Years later, I become part of a post-graduate learning program specializing in how to better understand Neurodevelopmental Disabilities and ways that families cope. I come to know more about these parents. I come to see how they exhibit resilience and a "whatever-it-takes" mentality with their lives often consumed with advocacy at the medical level and educational level. I vow to work to help those families with special needs in my future career so I can help those whose path I was almost on.

Why Resilience?

Resilience (or resiliency) is a person's ability to adapt and bounce back when things don't go as planned. Resilient people don't wallow or dwell on failures; they acknowledge the situation, learn from both the gifts and the mistakes, and then move forward.

With resilience, we surpass our own limitations—even under the most challenging circumstances. Caregivers seem to be torchbearers in this domain. As individuals, resiliency is present in our ability to overcome challenges of all kinds—trauma, tragedy, personal crises, daily life problems—and bounce back stronger, wiser, and more personally powerful.

Joy versus Happiness

Wisdom tells us that joy is intrinsic well-being at the foundational level. Its outward expression might include happiness, but happiness is a transient emotion. Finding joy in our caregiving work has a larger purpose as a consistent expression of well-being in guiding our choice—not simply random happiness or being in a jovial mood. Any caregiver worth their salt knows that will never happen. Joy is deeper, more foundational, more accepting and abiding. It is the quality that lets us release our attachment to results and go with the flow of life as it is—not as we wish it to be.

Does resilience relate to joy? Can joy actually be another fruit of our care?

When joy is the guiding light of our care, we still have bad days. Upsets occur, and we clean up messes just like we always did. Yet something has changed, and we are at ease with the process and at peace with life itself.

Once we have our own perspectives in order, we are able to shift the emphasis from challenge to choice and open to the graces that surround us even in the hardest of times. However, sometimes, when we reflect inwardly, we find joy is missing. And this needs addressing. Depression and anxiety are realities that affect caregivers everywhere. Recognizing them and taking steps to address them with your physician is important from a medical vantage point, but doing small things like adding joy into your habits and routine can be done now.

What if Joy Is Missing?

It's hard to think about joy when you are stressed—and in a family caregiving circumstance, finding joy might seem even more challenging, but it's not impossible and *must* be seen as a priority. Like stopping to watch a sunset. Sure, you could do other things, but savoring that view and just finding joy in the moment is probably what you need right then. If you find time to do things that bring you joy and refresh you—even if just for a few moments—you will be able to find a sense of ease coming to your caregiving. Remembering what you loved to do in the past (or still love doing but can't find time for) is an important way to revisit what can revitalize you in a time of need.

Caregivers often stop the daily practices that bring joy simply because they forget to do the small things that once helped them stabilize. They forget to put on the music. They forget to pull up the blinds. They forget to sing in the shower. They forget to take a deep breath.

Having a family member whose care often takes a priority in life may feel like an overwhelming experience, but there are things that can be done to bring back a renewed joy in yourself. Find those areas in life that do bring joy. Remove some things that are joyless if you can. Say no powerfully and mean it. This is not about disappointing others. It's about keeping the caregiver in a state of balance. It's very hard to take care of others when you are burned out (and the so-called candles are being burned at both ends).

The Prospect of Leadership as Caregivers

We often think of leaders as visible people who seek to claim the leader mantle, such as political figures or managers of companies or organizations. Many would hardly recognize themselves as the phenomenon of servant as leader in their role of a family caregiver, but it is worth exploring. Modeling good care for yourself is part of being a good caregiver. This ability to be creative while helping others (and ourselves) shows caregivers as the leaders we are. And we, caregivers, are in fact, leaders in our own right—it just has to be recognized and practiced. When we position family caregivers as leaders who have a vital contribution to make, we often see our fierce and protective nature step forward most clearly in difficult caregiving moments.

Gael's work with women-led initiatives brought her to the organization Athena International and the legacy of Martha Mayhood Mertz, author of *Becoming Athena, Eight Principles of Enlightened Leadership* (2009). Mertz outlines inter-connecting systems that family caregivers demonstrate every day. In leading us into what is now a global movement, she sees these actions "as ancient and organic as the social patterns established by women over centuries," and something that is as vital and timely as the innovation we need to lead us in the 21st. The eight pillars highlight the practices of authenticity and lifelong learning, courage and fierce advocacy, collaboration and relationship building, giving back, and celebrating those around us. In speaking to the aspect of building relationship, Mertz refers to Max De Pree, the successful former CEO of Herman Miller, Inc. Author, De Pree exudes wisdom in the book *Leadership Is An Art* (1989) saying that "leadership is more tribal than scientific, more a weaving of relationship than an amassing of information." With his uniquely different perspective on leadership, De Pree places emphasis upon building relationships, generating ideas, and staying attuned to your own value system. For this, leadership can begin in families and expand outward to corporations, all of which can be, "a place of healing.... a place where work becomes redemptive, where every person is included in her own terms. We know in our hearts that to be included is both beautiful and right." Mertz and De Pree offer a different view of how leadership is accessible and attainable. Often caregivers embody these attributes in their lives already, and yet may not claim the title.

Caregivers as Leaders with Resilient Habits

When James Kouzes and Barry Posner, researchers and creators of the Leadership Challenge, first set out to discover what great leaders actually do when they are at their personal best, they collected thousands of stories from ordinary people—the experiences people from all walks of life recalled when asked to think of a peak leadership experience. From this research they distilled several traits of exemplary leadership that all of their most resilient leaders shared. Gael and Sharon found striking similarities between leaders in Kouzes and Posner's research and todays family caregivers because both exhibit resilience and personal strength even in times of great stress. We set out to explore for ourselves some of the traits caregivers have when they have learned to thrive in the midst of overwhelm.

The term leadership might at first seem like a misnomer to assign to caregivers who appear to be living at their peak despite the challenges. To live a life with meaning and self-replenishing takes proactive planning and a positive perspective on the harder parts that come during times of crisis. Both Gael and Sharon recalled numerous individuals who have achieved a normalcy in the midst of ups and downs of caregiving and who still managed to grow in their own lives while caring for others.

Resilient caregivers tend to have beneficial patterns of behavior—regardless of their walk in life, their age, or their experience. We found these elements across the board, and they match exemplary leadership qualities to a tee: clarity, managing overwhelm, sharing the importance of what they are doing, addressing challenges head-on, and upholding dignity and respect in everything they do. Within a care team or a family group, having a state of mind that is resilient is contagious. And caregivers can guide the way with habits others can learn from.

Clearheadedness in the Midst of Crisis

It is often hard to be clearheaded in the midst of a crisis or even a long-term difficulty related to a family member, but it is possible to cultivate a sense of clarity. Caregivers who adopt the time and mental habits to attain clarity benefit from having a plan and intentions surrounding what they do. The ability to be clear with their actions allows them to move forward in a way that aligns with what they believe in.

Moving a parent out of their home or having a child change educational environments that better support their goals can be hard when the status quo feels easier. Getting buy-in for family members who are reluctant to embrace a hard change (even if they will benefit later on) requires convincing from someone whose advice aligns with the values of the family. With clarity of mind, a caregiver is able to get Mom or Dad moved to a safe residential care facility. For a child whose educational goals are not being met, then changing schools or adjusting classes or adding structure can mean readjustment but the longer term benefit that is in mind must be part of the rationale for making the change. There are benefits to working through these big or small course corrections in life but meeting the end goal will require patience. It is important to recognize successes and celebrate how clarity made those actions possible.

Managing Overwhelming Situations

Caregivers are no strangers to moments of crisis and endless multitasking. Those caregivers know how to stay grounded in the midst of chaos and are able to thrive when others might be overwhelmed or find their actions stifled. Caregivers who are resilient seem to be experts at setting interim goals and executing actions to avoid drowning in complexity.

Learning the diagnosis that is scary or overwhelming can make a person feel overcome—even their family may feel that life is never going to be the same. A caregiver can employ steps to help the family member know that action plans and a sense of hopefulness are tools in the journey forward—like a tether to pull through a dark or unclear path. Setting an example of calm and hopefulness allows caregiving leaders to guide the steps with more ease than those who don't have those habits in clear sight.

Sharing the Importance of Care

Another hallmark of leadership that some caregivers have is an ability to show how their care of another is part of a larger worldview. Sharon knew of a caregiving couple named Mike and Karen who had a family member with Lewy Body Dementia. As they scoured their local area for like minded caregivers coming together to help each other, they found it lacking. They decided to start a local caregiver support group – again, only devoted to caregivers—and created a community that helped them and others. In these meetings, the couple were able to show caregivers in their area that there is a place to be present, honored for this role, and learn ways to succeed in places where the difference could be felt. This couple having started a support group didn't do so with leadership on their mind, it just met a larger objective of finding people to commiserate with. Regardless of the motive, there was a sense that there was value in caring for others—whether it be family or the wider community.

In the workplace, we see examples of how this shared vision of support in a time of care is growing when Human Resources helps caregivers with accommodations or time off. We also see this influence growing in policy and governmental organizations who lobby for caregivers. Over time, individuals and organizations sharing this vision for supported family care will make a difference. Caregivers recognize that this is a communal effort to attain a future that involves health care, home care, finance, work, and an adequate network of services. We all grow if we share this common vision, and it will breathe life into our intentions for care by sharing the possibilities we see with others.

Addressing Challenges Head-On

Caregivers who are leaders are those who, in the midst of care, bump up against frustrations and reflect upon how this may be an opportunity to change the status quo. It doesn't always mean caregivers who address challenges head-on are overbearing or taking on fights. Rather, they see challenging moments as chances to improve through experimenting and taking risks with new approaches. These caregivers see that taking risks involves a certain number of mistakes and failures. It also creates valuable learning opportunities that they and others can benefit from.

We all know the ups and downs of trying—and sometimes failing—to get things done. We all experience the carousel of adapting to change in the simplest of moments, and taking on a boldness that might be required to handle a situation head-on can be challenging. However, that is where our resilience lies. Those who are unafraid to ask or persist when an inclusive ramp can be added for wheelchairs or appealing a no from an insurance company or provider. Hard work to get a yes can make a difference.

When a caregiver makes a difference with getting their voice heard their lives get better and people see it in their lives. There are even times when the internet sometimes "hears" a story of a caregiver's advocacy having a big payoff— occasionally this type of news goes viral. Just think of those occasions you read about on the news or your social media stream when you hear of a family taking on an insurance company and managing to override rejection and not only get their medicine or other such care covered, but their role also had the impact of making a treatment available for others in that situation. Powerful advocacy can start small but radiate outward. Remember that paradigms are meant to shift. It just takes people with the nerve to try to move the need. Caregiving leaders are those people.

Upholding Dignity and Respect

In all domains where exemplary caregivers act as leaders, they know what it is like to have their situation (or that of the family member cared for) minimized. There might be those without courage in the moment to accept that their poor treatment is acceptable. Caregivers who are able to voice their commitment to dignity and respect, even in situations where they are being disrespected, are able to bring their concerns to light professionally and elucidate how treating each individual with respect can be a mutually beneficial experience for all.

It is often the caregiver who sustains the dignity of their loved one, and they can model this for all in the presence of medical staff, helpers, and other family members. As we create our systems of communication and support, we are in essence enabling others to act to uphold dignity and respect as a powerful tool for influence. It is important to know that to keep hope and determination alive, caregiving leaders routinely recognize the contributions that all individuals make and make people feel like heroes. This is the essence of respect toward those we live with and work with, and it requires a habit of the heart that can be cultivated and maintained.

When we embrace clarity, managing overwhelm, sharing a vision, managing challenges head-on, and upholding dignity and respect, we create a framework for resilience in the process of our own caregiving leadership—whether we are focused on one beloved family member, participating with an entire team of staff and family members, or managing groups. When living out a shared mission, it is always a good thing to recognize and celebrate a consistent validation of the wins. What are yours?

today
I CHOOSE
joy

Chapter 10 Reflections: The Power of Resilience

- What are the ways you are committed to celebrating your wins, large and small? Begin a journal or a simple daily checklist to validate each and every one. No win is too small. Make this a daily practice. You'll be amazed at the results.
- How do you make time for the practice of cultivating joy?
- What is your dream for improving the status of caregivers? Do you see yourself having a role?
- Do you plan for things that you look forward to? Is there a museum exhibit in future months you can book tickets for now? Can you plan to carve out time each week or month for the things you want to do? Get out a calendar, think it through, pencil it in, and get yourself going in the direction of committing to a plan that upholds your own self-care as a priority.

Notes to Myself

CHAPTER 11
Concluding Thoughts

In three words, I can sum up everything
I've learned about life: it goes on.
—Robert Frost

The Journey Home: Gael

February 1

Mom is dying. Taking care of Dad was her primary job, and she put so much of her life effort there. Grief and ease, ease and grief, they are best friends on the playground of emotion, though they seem an unlikely pair. Grief and ease play a balancing act with the seesaw of our hearts, and today, I see why this death—her death—is so different from the one we experienced three years before with my dad.

For every kind of help we four children gave through his decline, the grief of loss was balanced with the energy of ease. We were helping Mom, and she would have done it all by herself if she could have.

Ease gives a reward, recognized or not, and it helps everyone carry on through the challenge of loss. There is no such balance now.

This grief is different—more one-sided. Do I have many more days with her? More weeks? Is it mere hours to hold her near? My heart feels such sadness, and faith is what I seek.

February 6

Mom was started morphine by mouth twice a day earlier this week. Today it was upped to three times. I have come to spend the afternoon with her, but she is no longer waiting by the door with the relief of a puppy who is going out. Today, she is lying in bed, eyes unfocused. The hospice nurse says her condition is likely more than the morphine—the tumor is probably making its way into her brain in the same way we see it growing outward, becoming a hostile takeover on the side of her face. She had been asking, "Why am I still here? I'm so ready to see Jesus."

You must be here for me, I think, For us!

Yes, you are here for each of us, and for the experience we all have when the only thing left to share is love.

February 8

Like a whisper, Mom is gone.

How Caregiving Is a Tribute to Those We Love

People can thrive when caring for others. For some, it may come easier, and others have to work at it. Caregiving is in many ways a tribute, and it may help strengthen connections to a loved one. You can find joy or fulfillment in looking after others, but there is a strain that needs to be addressed and efforts taken to recharge or it can become overwhelming. The power lies in striking a balance.

Finding Stability as Uncertainty Continues

Making sense of caregiving is hardest to do when the challenges are at their peak. To attempt to find a new way forward with new habits and a focus on doing things differently is best done with detachment. Although it is best to reflect on a past caregiving experience when things calm and a routine is established, not every family member cared for has a condition that will improve or stabilize. Cancer might go into remission, giving one the feeling of relief. but it can be marred by intense levels of uncertainty and anxiety because a hallmark of cancer is that it may return months or even years later. A medical diagnosis might be found after years of looking. But once the diagnosis is clear, it can seem hard to find that challenges exist even after knowing the condition or disease that was a mystery before.

As a caregiver with ongoing challenges that may peak or subside over time, it is all the more important to create a sense of calm in the chaos of life to regain perspective of the experience. Revisiting reflective questions offered in chapters such as "Caring for the Caregiver" or "Something's Gotta Give" will allow you to steady yourself as you find ways to cope that are helpful and not hurtful to you or others. In addition, speaking to others in a support group for caregivers as well as keeping channels of communication open with those in your current social group will help you maintain a sense of normalcy. Avoiding isolation will prove useful to relieve stress though interacting with others and finding joy in a challenging circumstance.

Anticipatory Grief and Post-Death Bereavement

Those with caregiving concerns that don't involve a terminal illness may not come to see a different type of caregiving shift— that which happens when caregiving comes to an end. Many who care for family members who are dying see the decline and begin to experience what experts call *anticipatory grief.* This term describes the mourning by both patient and caregivers and others close to the person cared for before the actual death. This notion was first written about by psychiatrist Erich Lindemann in 1944 and was explored by psychiatrist Knight Aldrich in 1963. Although we can anticipate what will come next, it is a confusing and sad experience when death occurs.

Author and psychotherapist, Judy Tatelbaum, writes in her book *The Courage to Grief* (1990) her thoughts related to how caregivers respond to the experience of ending their time of care with death:

> There is a natural sense of loss when the need for our caregiving is over. We must often face the double sorrow of losing a loved one and our purpose or role in their lives. The aftermath can be a very difficult time that leaves us feeling lost, lonely, and useless. We may not feel grounded without that important function of taking care of another in our lives. Our direction may feel unclear. The future may look bleak or even empty.

Although these feelings may seem overwhelming, talking with others and getting a sense of perspective helps hope to be found and a way forward out of sadness.

Seeking support from bereavement groups is a helpful way to feel the sadness among those feeling similarly. It may also help cope with emotions that are inevitable after death. For many, there is grief from the loved one's loss, and there can be an acute sense of the end of the role or identity that caregiving has played. For some, caregiving was a way of life to attend to someone else, and it can create a need to build a new life. For some, it is possible to feel guilt or relief that many intense caregiving duties are over. None of these feelings are bad or wrong.

Perspective

The ability to step back from care allows for new thinking with insights perhaps not before considered. In Sharon's *Bridging Generations* study (2013), caregivers were asked about their time of caregiving. Many participants looked back on a challenging time in the past and found they regained a different perspective of their care for others. Mario learned a great deal about himself and those for whom he was caring. "I recognized that I was taking care of my family and me and I was giving back." Your own perspective of family caregiving is unique to your situation. Find a way to honor that by allowing time to reflect on the time of care in the past or in the present.

Implications for the Future of Caregivers

Caregiving is a labor of love, and for many primary caregivers, this becomes more than a typical presence in a family. The National Institutes of Health estimated in 2015 that 43 million American adults provide this type of unpaid care for someone in their family with a serious health condition, and it will only grow in years to come. As the elderly population continues to grow nationwide, so will the need for family caregivers and those who are paid or unpaid that family members can delegate to.

Those in a family who take on the caregiving role without any training can put themselves at risk for stressors that effect other parts of their lives if they don't take precautionary measures that restore them and ensure that they stay healthy. Unfortunately, many caregivers don't. Experts at the National Institutes of Health created a report in their series titled "NIH News in Health" and a December 2017 special edition focused upon caregivers. In "Coping with Caregiving: Take Care of Yourself While Caring for Others," it stated that caregivers may be less likely to meet their own needs – like filling their own doctor's prescription or they may forget or delay things like preventative screenings which are helpful in early detection of health problems that are present or coming later on. Also in this newsletter, Dr. Erin Kent, an NIH expert on cancer caregiving, stated, "Caregivers also tend to report lower levels of physical activity, poorer nutrition, and poorer sleep or sleep disturbance." These types of lapses in health can worsen over time, and caregivers will pay a price for sacrificing their own health in a time of caring for another.

Learning how to be attentive to those you care for and yourself are sometimes skills that can falter. If you need to revisit ways to advocate or advance your own self-care, mark the pages in this book and reread the chapters that apply. Search online for subject matter that helps you continue to find your voice and act powerfully in your own best interest as the center of your family member's care team. Your role is valuable and needed—even if the acknowledgment from others is not explicit.

The Future (Owning the Role)

The future will change as more caregivers own the role and make their voices heard. Organizations with policy objectives like Caring Across Generations and their ideas staked for the #Caring Majority are making inroads with leadership in many states. As their platform states, they are creating action to help families for whom caregiving is a way of life due to medical issues. Their website (CaringAcross. org) states,

> Planning for long-term care is not just personal— it's political. We're working to bring our caregiving infrastructure into the twenty-first century, so everyone can age and care with dignity. We're living longer than ever before, and we can't always expect family members to give up everything to care for us. It's time for policies that support both seniors and people with disabilities who need care, and the families and professional caregivers who care for them.

With Caring Across generations and other organizations like Family Caregivers Alliance helping to advance caregivers' visibility and needs, support for families is getting the platform it needs to improve the future for caregivers.

A Multigenerational Understanding of Care

Participants in the Bridging Generation (2012) study explained how their understanding of care changed over time. A few remarked on how they witnessed grandparents being cared for by a mother or father as kids and had never retraced these actions into the present time—until they tread the familiar steps of caring for an elder in the present day. Mike, a study participant, was fearful of his own grandmother's death as a teenager. He explained how his own stepchildren are more aware and empathic after seeing family care in their home. He saw how his granddaughter who was a witness to care came to see caregiving as just how their family looks after each other, and it was not seen as a burdensome or scary task. Even as his mother-in-law's condition brought challenges from her dementia, there were memorable times of three generations laughing at the dinner table. He described how memorable it was to wash dishes slowly to savor the togetherness. He won't forget those small meaningful moments—and neither will his granddaughter.

From one generation to the next, care will be passed on like a torch. Yet with an openness and ability to take on meaning amidst challenges, the time of caring for a family member is not to be avoided. Even if a comment is made or a resentful feeling is uttered, feelings can change over time. Perspective can take root. A richness of love can be powerfully gleaned during even the most challenging family crises, but you have to work to attain it. For those who are steeped in chaos and not able to find joy, this richness is not something they will be able to see. But by making sound choices and having solid boundaries, a time of care can be fruitful and powerful for all involved.

Make a general rule for yourself to never give up on fostering your own resilience. Take the wisdom of having learned self-care into your life and own it even after caregiving ends. Be a mentor to others so they don't see self-care as selfish. Role-modeling for others is a vital part of showing how family care can be done in a way that shows love of self and of others without sacrifice.

Below is an excerpt from a college essay written by Sharon's daughter, Erika, who was witness to family care in her life as a young child. Her story shows how our care crosses generations, and it touches on the hard emotions, which at first involved jealousy and resentment from a young child's memory and ended in grace and acceptance from a more mature young adult perspective. Erika's story shows the value of emotions and how these feelings can evolve with maturity in our kids' lives. As caregivers, we have to remember to be leaders for our families, and we need to be open and aware of the difficult challenges of care and the beauty and poignancy that can arise in care or after care.

What They Left Behind: Erika Marts,
Written at Age Seventeen

His leather Air Force jacket lay in a box in the garage. Her wedding dress is folded and tucked away in the attic. I see their pictures scattered among the house, and the faint recollections I have of them come back to me. I never knew them as my parents did. My parents spoke about them with admiration, yet with evident sadness; when I had come into the world, they were only a shell of what they had been. Their battle with Parkinson's Disease had crippled them both. My grandfather was a daredevil pilot in Vietnam; those around him saw him as brave and honorable. I had been scared to talk to him; sitting in his wheelchair in his frail state, I had never given him a chance to tell me about his experiences in the sky. Moments where he was suspended between the world he was familiar with, yet not floating in space. My grandmother was a teacher, with wild curls; although we were the only two in my family with the trait, I never tried to see what else we had in common.

At nine years old, I resented the space their wheelchairs took up in the car. I questioned the importance of my mother's actions as she went out of her way in order to have her parents present at every family dinner and Sunday morning church service. At their funeral services, sitting in the church with the thick smell of incense and a drowning presence of grief and sadness, I did not see the ways that their presence had impacted the aspects of my identity that I am most proud of.

Thinking about my life without them as I had when I was nine was vastly different than reality; their absence was more prominent than I could have expected. We no longer sat as a family around the couch, laughing as my father repeated a joke he had been told by my grandfather when they had first met. We didn't make hamburgers on Christmas Eve, as we had every year they spent Christmas with us. I had never understood how much I savored the moments in which our family was together; as we laughed, we were united in a way that we would never fully be again.

I realize now that the beliefs and values I feel so strongly about were not just important to my parents; as my mother raised me, she passed on what her parents taught her. Our weekly attendance at a church service, which had begun with my grandparents' devout conviction in religion, reinforced the importance of faith in my life. My beliefs provide me with solace, knowing that they guide and protect me each day.

My grandparents' legacy lies within my sense of what is right and wrong, my unconditional love for my family, and my strong will. I know that life is not centered around receiving; this was exemplified in my mother's values and the love she had for her parents as she cared for them when she knew they needed it. In the same way, I know that I am willing to help those who I care for eternally. The remorse I feel as I have reached a higher level of emotional understanding has reinforced the importance of cherishing those who are in my life now. As resentment has turned into regret, I am reminded of how much I value forgiveness. I hope to live without indignation or contempt for others. Just as I was taught through my grandparents' actions, I will continue to value *hard work, perseverance during hardship, and selflessness. Today, I am thankful for the time I had with them. Even though I can never physically tell them that I appreciate them, I can only hope that I make them proud.*

Chapter 11 Final Reflection: Concluding Thoughts

Taken together, the compilation of stories we tell ourselves about caregiving becomes a narrative. In looking back at your own reflections, note what you've gained in writing your way into a deeper understanding of your own journey.

Best wishes to you, the caregiver, as you continue to do what you do.

Notes to Myself

Notes to Myself

Photo by Bob Alba

Gael Chiarella Alba and Sharon Zint Marts,
Authors of *Fruits of Care: A User's Guide to Family Caregiving*

For more inspiration, visit the authors' websites
at SharonMarts.com and Gael.info